Advance Praise for *One With the Tiger*

"In this engaging volume, essayist Church uses the story of David Villalobos's 2012 jump into the Bronx Zoo's tiger cage to launch a broader discussion on the connections people try to forge with animals—and the blurry line between humans and beasts . . . Readers expecting a narrow examination of Villalobos's tiger encounter at the zoo will be rewarded instead with Church's insightful exploration of human infatuation with nonhuman animals."

—*Publishers Weekly*

"An exploration of the fascination with the 'savage and the wild inside' us, which fuels the human desire to 'to get intimately close to apex predators'. . . a powerfully written attention-grabber."

—*Kirkus*

"From the iron of a zoo cage's bars to the expanse of our nation's national parks, *One With the Tiger* examines the spaces in which humans contain animals, and how those acts of containment often fail. Church is a classically essayistic observer—curious, haunted, self-deprecating—and it's through this lens that we're confronted with stories of infamous animal attacks, pop culture icons, and the author's own longing to inch forward as a bear approaches. In this marvelous collection, Church seems to write his consciousness directly onto the page, and in it we can see an entire civilization's clumsy, sometimes desperate, attempts to understand our relationship to the wild."

—Kristen Radtke, author of *Imagine Wanting Only This*

ONE WITH THE TIGER

ONE WITH THE TIGER

*Sublime and Violent Encounters
Between Humans and Animals*

STEVEN CHURCH

SOFT SKULL PRESS
AN IMPRINT OF COUNTERPOINT

Library of Congress Cataloging-in-Publication Data:
Library of Congress Cataloging-in-Publication Data

Names: Church, Steven, author.
Title: One with the tiger : sublime and violent encounters between humans and
  animals / Steven Church.
Description: Berkeley, CA : Soft Skull Press, 2016. | Includes
  bibliographical references and index.
Identifiers: LCCN 2016040116 | ISBN 9781593766504 (alk. paper)
Subjects: LCSH: Animal attacks—Anecdotes. | Human-animal relationships.
Classification: LCC QL100.5 .C58 2016 | DDC 591.5/3—dc23
LC record available at https://lccn.loc.gov/2016040116

Cover design by Faceout Studio
Interior design by www.DominiDragoone.com

SOFT SKULL PRESS
An imprint of Counterpoint
2560 Ninth Street, Suite 318
Berkeley, CA 94710
www.softskull.com

Printed in the United States of America
Distributed by Publishers Group West

10 9 8 7 6 5 4 3 2 1

For Andrea

TABLE OF CONTENTS

DAVID'S LEAP

I have a huge and savage conscience that
won't let me get away with things.

— OCTAVIA BUTLER

On Sept. 21, 2012, twenty-five-year-old David Villalobos purchased a pass for the Bronx Zoo and a $5 ticket for a ride on the Bengali Express Monorail. The ride, built in 1977, promises on the Zoo's website to take the visitor "above mud wallows, pastures, forests, and riverbanks to the heart of Wild Asia." After leaving the station, David would've first crossed the mud wallows of the Bronx River, a shallow and polluted urban waterway, before making his way quickly into said heart of Wild Asia.

In September 2014 when I purchased my own ticket to ride, the conductor and tour guide, Devin, a twentysomething guy in khaki pants and a retail haircut, told us that the Bronx River was home to a pair of beavers, the first wild beavers spotted in the river in over a hundred years. One of them, Devin told us, was named "Jose."

"His partner, the other beaver," he said, "voted on by the public, is named 'Justine Beaver.'"

Everyone laughed at Devin's joke and stared down, searching for the beavers below, hoping to catch site of the local native celebrities. The monorail's cars are built so that you only sit on one side, facing the left of the track instead of toward the front of the train; and though they have roofs with translucent skylights, the viewing areas open to the elements, bordered only by a short railing of metal tubes. It can make for a great show as you skirt the perimeter in what feels like a moving couch or section of sports bleachers; but on this day, we saw no celebrity beavers.

Devin drove the train, talked into a microphone, and played some prerecorded narratives about the animals and the zoo's conservation efforts. The track circles the perimeter of the "Wild Asia" exhibits, and it feels like you're waiting for a show of some kind. On his own ride, David Villalobos positioned himself in the last car, far from the conductor, and he listened patiently along with the rest of the visitors, waiting for his chance.

The exhibits in Wild Asia mostly consist of various species of deer and cattle, all of which look only slightly different in size and shape from the deer and cattle that we all know. If he had been looking for them, David might have seen spotted deer and Taiwanese Formosan deer and one called a "hog deer." There are also wild horses, massive buffalo-esque cattle, hairless jungle pigs, and rock-climbing goats. But the king of the mountain—so to speak—and the animal most of us were waiting to see, was the Siberian tiger.

The day Villalobos visited in 2012, a 400-pound Siberian tiger named Bashuta was "on exhibit." The tigers are the only apex predators in Wild Asia, and their habitat is separated in part from the monorail and the traffic on the nearby Bronx Expressway by a tall wood-pike fence that looks like something out of the movie *King Kong*. This is one of two tiger habitats in the zoo, the other being a place called Tiger Mountain, where you view the cats not from above but instead through thick panels of glass.

As the monorail tram left behind Cali the rhinoceros and curved past the fence, the open side of the tiger paddock slowly would've revealed itself, almost like a curtain parting, and a grand green stage, set for drama, open for the existential business of witness. The tramcars ran right along the top of the chain-link boundary fence, sixteen feet up in the air, and David would've been able to look directly down into the grassy hillside habitat.

On the day I visited, the weather was perfect—sunny and about eighty degrees with a slight breeze—and the paddock looked like some kind of idyllic picnic spot or forested campsite. The air was cool and seemed like it had been freshly washed. Three tigers live in Wild Asia, but because they are solitary and fiercely territorial, only one of them is on exhibit at any given time. Devin told me that one of the tigers likes to lay up against the fence, just beneath where the tram passes and is, thus, hard to see without leaning out over the edge of the car.

For my visit, a Siberian tiger named Yuri lounged on the hillside in the sun, raising his head to consider us as we passed. It took me a while to spot him at first, but I followed where Devin pointed. Strangely, it was easier to see the tiger if you didn't look directly at him. Yuri was enormous and regal, a striped king reclined on a blanket of green; not hiding, he was still somehow camouflaged against the background, his stripes creating a veil of color and shade.

Three young boys traveling in the same car with me drummed their feet against the fiberglass benches out of either boredom or excitement as their nannies and grannies squealed and pointed at Yuri.

Though Yuri seemed to be napping like a common house cat, he was also menacing in a way that's hard to articulate. Even from a significant distance he seemed dangerous. Something about his body—the size of his paws and his head, in particular—broadcast power and violence. Yuri was fun to look at, and it sent a little tingle up my spine to see him there in his habitat. But I was fine with seeing him from afar. As an object in a cage. Contained and controlled. I didn't need to be much closer.

The tiger habitat is shaped like a shorter, fatter football, and the backside of it, at the top of the hill, shares a chain-link fence boundary with exotic cattle and deer habitats—which I imagine must be, for the tigers, like staring outside your prison bars to see not just freedom but also a buffet of all the food you could possibly eat; and our train appeared regularly like a line of sushi boats floating just out of reach.

David Villalobos didn't care about the Formosan deer or the hog deer or the wild horses with their short, bristly manes. He positioned himself in the very last car, and he waited for the guide to take them past the jungle pig and Cali the rhino and the wood-pike fence, until the car crept along the top of the tiger enclosure. He'd planned this. He knew exactly what he was doing. He had to have taken the ride a few times, gauging the distance, timing the tour and the movement of the tram. I imagine the zoo employees had come to know him, to recognize his face in line. Maybe they'd started to wonder why he came so often.

David knew what was about to happen: the guide slowed the cars, trying to spot Bashuta in the brush, and pointing her out to the other visitors. David waited for his cue, for the opportunity. As his car pulled up alongside the paddock, Villalobos suddenly stood up. He climbed up on the edge of the car, reaching up and bracing himself against the roof with his palms. He balanced briefly on the railing, rolling on the balls of his feet to get his balance right. Then he leaped, clearing four strands of barbed wire sixteen feet down into the tiger's cage.

David landed on all fours, "like a cat," he would later brag, then crumpled and rolled to the side.

People on the train gasped, pointed, and screamed.

Imagine the terror of witnesses. Think of how that moment must have shimmered and buzzed, electric with fear. I've tried to imagine David's leap, perhaps as a way to get closer to understanding it. I'd come to New York to try and retrace his steps. I'd come to the zoo to find David.

"He jumped!" someone called ahead to the tram conductor. "Stop the train! That man. He jumped down there! Into the cage!"

Was this part of the tour? Is this real? Is it some kind of show? Already the question rises, as if released from the earth upon impact, like a cloud of dust: *Why would he do that?* And that question would linger, still floating in the air two years later when I visited.

David looked up at them and saw their mouths stretched into dramatic "O's," their arms and fingers extended, pointing, the other hand covering the eyes of a child or waving frantically, telling him to run, to get away—like people do when they're watching a horror movie and the actors are doing things you know are going to get them killed; and perhaps David smiled and waved back at the people above, his witnesses and his audience. Or perhaps he was so focused on his mission that he didn't even acknowledge the spectators and their commotion, and he simply gathered himself to stand and face the tiger.

////

BASHUTA, WHO DESPITE HER fairly domestic existence in Wild Asia still possessed all the instincts of an apex predator, made her way quickly over to David. I see him there, in the long slow seconds of those first moments in the cage, turning and smiling, his thick shock of black hair standing out against the green background, his eyebrows arching into eternity. There is no sound to these images. Bashuta's huge paws pad silently on the grass, covering the ground between them in seconds. And if we pause the scene, you will notice that David is a handsome young man. Tall and slender, with those thick black eyebrows and high cheekbones—he could be a model, he could be your boyfriend or classmate, maybe just the guy you see every day on the subway into the city and you think to yourself that he looks familiar or famous or otherwise interesting. But in those ecstatic moments in the tiger's cage, the sun dapples his face and he holds his arms out to the animal, as if to embrace Bashuta and, for a second, the scene *is* beautiful. Time slows down, filtered through the thick lens of memory. Each witness to David's leap must, inevitably, have his own story, his own burden of that day.

In the moment, only one thing is true: David wanted to touch the tiger. And he would touch her. She would touch him. She would touch him deeply. She would ravage his foot; puncture a lung, and more. Tigers typically kill their prey by biting the neck, snapping the bones and puncturing vital arteries before dragging the body off to a secluded spot where they can feed. David spent close to ten minutes alone with Bashuta. Ten long minutes in the cage. And he did indeed suffer several broken bones, at least a few of which may have been the result of his leap from the tram and his apparently less-than-catlike landing.

It must have taken a while for word of David's leap to spread in waves through the tramcar, up to the driver, and eventually to zoo officials. The tram operator, probably a kid just like Devin, who most certainly radioed ahead to tell them what had happened, would've had no choice but to leave Villalobos there and make his way past the goats and hog deer, past Wayne the Red Panda, the Pygmy Deer, and quickly back to the station. He had a responsibility to the other visitors. He had to get them out of there. The zoo certainly didn't want a tram full of visitors to witness a man being eaten by a tiger; and that's exactly what everyone expected to happen. David didn't stand a chance. He was dead meat as soon as he landed in that cage.

Zookeepers sprinted into action to try and rescue David, following a response protocol that they'd practiced but were rarely called upon to perform. They had to act fast and fearlessly. Rushing to the scene, one zookeeper blasted a fire extinguisher into the cage (an oddly common method of intervention in such situations), frightening Bashuta away, as another zookeeper instructed Villalobos who, to everyone's surprise, was still alive, to roll toward them.

David *was* bitten and clawed and dragged around by his foot, and he suffered numerous bite wounds; but Bashuta did not break his neck and did not kill him. For whatever reason, the tiger displayed an unexpected and unpredictable level of restraint and patience—behavior that looks, in retrospect, a lot like mercy.

Or perhaps, as one zoo employee told me, "He was just lucky that tiger wasn't hungry."

In the second-day news stories, Villalobos was described by his attorney Corey Sokoler as "very intelligent" and "very caring," and reports surfaced that David had told the responding New York City Police officer, Detective Matthew McCrossen, that as the tiger mauled him, he'd stroked Bashuta's face, petting the beast like a common house cat.

David believed he'd forged a bond with the tiger, and that he'd crossed over and developed a connection that was hard to describe. Perhaps David imagined himself someone who lived between the civilized world and wild nature. Perhaps he believed he'd succeeded in bridging the divide between human and animal, that he had crossed over to their side, if only for a second.

Maybe he was right. Maybe I was a little jealous. Maybe that's why I traced his trip to the zoo. Maybe I just couldn't get his story out of my head. Ever since I read the first reports of David's leap, I'd been somewhat obsessed, losing myself down rabbit holes of research into similar stories.

David was arrested and charged with felony trespassing. His parents blamed Adderall for his behavior; and perhaps his mind *had* gone a bit wild and unruly, a bit savage, but David told officials that *his* leap wasn't caused by drugs and wasn't a suicide attempt. He wasn't depressed or delusional. He wasn't even really trespassing.

He was going home.

David told anyone who would listen, "I was testing my natural fear. You would not understand. It is a spiritual thing, I wanted to be at one with the tiger."

////

WHEN DAVID VILLALOBOS JUMPED from Bengali Express Monorail into Bashuta's cage, he also leaped straight into my consciousness. Something about his story cracked open a well of curiosity that bordered on obsession. I followed all the initial news reports and the follow-up stories about the incident. I couldn't get enough. I wanted all the information, all the facts and fluctuations, anything I could read about it. I wanted to get inside the story somehow. I'd heard of people leaping into cages with apex predators. But most of the time they don't survive to tell a story. David survived only to be vilified and mocked, publicly indicted as deranged and suicidal. But this story wasn't his story, or the whole story. Aside from his statements to the responding officers and his boastful claim, landing "like a cat," I could find few words from David himself. But his leap seemed to speak to me in ways that were hard to understand.

It seemed too easy to write him off as crazy. Even retelling the story of his leap here seems ridiculous, even comical or absurd, like a tale I trot out at parties to get gasps or laughs. *Who does such a thing?*

As is often the case, such questions become vessels for my own explorations. I wanted to understand the thinking of such savage and unruly minds. I wanted to get close to the subjectivity of people who push the boundaries between human and animal, who come close to crossing over; and I wanted to understand what drives someone like David Villalobos to make the leap, to thrust oneself into an encounter with an apex predator.

What did he mean by calling it a "spiritual thing"?

I wanted to understand what ecstasy exists, what promise of spiritual connection imbues such encounters and how it can seduce someone into risking his life. But I guess I'm also trying to understand my own interest and compulsion to come close to this experience, my lifelong desire to inhabit these tales of survival in the face of animal savagery, as well as the larger pop culture embrace of these stories. Why do such stories persist, and why do we persist in loving them?

////

PERHAPS SOME OF MY odd obsession started (or was rekindled) in the Fall of 2007, when I was contacted by a former student, an instructor in the Mass Communications and Journalism Department at the university where I teach. He presented me with an odd request.

He asked if I'd be willing to play-act the role of a bear attack victim for his beginning reporting class. He explained that there would be a mock press conference where his students would ask me questions. He explained that I just had to pretend that I'd survived a bear attack.

That was all.

"Just a little role-playing, a little acting," he promised.

WITH THE EXCEPTION OF my kindergarten non-speaking role as a munchkin in the sixth-grade production of the *The Wizard of Oz*, I'd never done *any* acting; and I knew next to nothing about pretending to be someone else. But I said yes anyway and began preparing for my role. I was told the students would be asking me questions for a "second-day," or follow-up story, and that I should be prepared to answer them from the perspective of the survivor. It dawned on me, vaguely at first, that I would be expected to offer a convincing facsimile of a bear attack victim. It dawned on me, quite clearly and quickly, that I was in *way* over my head and had a lot of work to do.

The Oscar award–winning star of the 2015 film *The Revenant*, Leonardo DiCaprio is famous for his dedication to the craft of pretending to be someone else. He works extremely hard to become another person. As a "method actor," DiCaprio tries to immerse himself as much as possible in the subjective reality of his characters. For *The Revenant*, this meant that DiCaprio did everything he could to *become* the single father, hunter, trapper, guide, and bear-attack survivor, Hugh Glass, embodying the character in a way that many have called "masterful." It was a performance that required a lot of physical sacrifice. And a lot of grunting, festering wounds, and visible suffering. He not only camped out in subzero temperatures and had to repeatedly dive into an ice-cold river wearing a bearskin coat, he also ate raw bison liver and—according to some (perhaps apocryphal) reports—he even slept inside a dead horse.

DiCaprio himself has said many times that he and the entire cast regularly risked hypothermia while filming in temperatures as cold as −40 degrees Fahrenheit. Hands became numb. Equipment froze up and stopped working. DiCaprio did not, however, subject himself to an actual bear attack—even if their film shoots in the Canadian Rockies were occasionally threatened by the presence of wild bears—but you get the sense, watching the film, that were Leo called upon to answer questions from a class of journalism students about what it was like to be attacked by a bear, he could answer them honestly and exhaustively. He could tell them *exactly* what it felt like, and you would believe every word he said.

For my acting gig as an attack victim, the students' mission was to uncover new details or new angles to my story, to find new meaning in events that had already been reported on; and my job was to both convince them that I'd actually been savaged by a grizzly bear, but also to confuse or derail them. But it now occurs to me that their assignment, their true mission, was perhaps my own as well.

My somewhat obsessive reading of attack stories, my imagining the horror and savagery, my desire to understand why David Villalobos would leap into a tiger cage, or why anyone would put oneself in close proximity to an apex predator, is, I suppose, driven in large part by my desire to discover new meaning beneath the surface details of these attack stories, to try and get inside the subjectivity of the experience. I wanted to "method act" my way closer to understanding.

The now-infamous bear attack sequence in *The Revenant* takes about five minutes of screen time. It begins slowly as DiCaprio's character Glass sets off alone in the forest, scouting a trail for the hunting party he's guiding. It's early morning, just after dawn, and the light is perfect. Trees tower overhead, swaying silver and black against a gray-blue sky. The sound here goes ambient and, at first, all you hear are twittering birds, the staccato knocking of woodpecker, and the white noise of wind blowing through the treetops. Glass's boots crunch through the brush.

He hears something, maybe. You're not sure at first what it could be. The group was being tracked and hunted by a Pawnee chief searching for his kidnapped daughter, and they were on the run. Perhaps they've been found.

Glass pulls his hood down and stuffs a wad of chewing tobacco in his lip. There's something out there. But you can't see anything. All you hear is Glass's breathing.

Or is that something else breathing?

You hear this deeper and more guttural sound, a huffing breath that doesn't seem human. It sounds bigger, more animal, and it pulses beneath the images on screen like a rumbling bass line that builds and builds. You realize subconsciously that you're listening to the bear. Your first fear is conditioned by this sound, by the bear's amplified breath. And it is a deep fear, one that you feel in your own chest. There is no music, no soundtrack, just the bear breathing.

Two mewling bear cubs appear onscreen, scrambling through the dense undergrowth. Glass raises his rifle; and just as he turns around to look for the mother bear, we see her, over his right shoulder. She rises up on her hind legs, bellows an angry cry, and looks straight at Glass, who stands between her and her cubs. She drops to the ground and charges fast in a mad growling rush. Glass doesn't even have time to turn around before he is slammed into a tree. The massive bear, weighing at least 400 pounds, rolls over Glass like a quivering wave of brown fur and teeth and claws; and the noise, the huffing and growling, the screaming, washes over you, pins you to your chair, and then recedes, leaving pools of silence where you know more danger is lurking.

At one point in the midst of the attack, the bear, standing on DiCaprio's body, leans in close to his face, almost nuzzling him and sniffing at his neck. Two or three times the bear pauses for these moments of odd intimacy that look a lot to you like mercy mixed with curiosity; and the second time, she actually uses her head to roll DiCaprio over on his back before licking at his bloodied face in an almost tender way.

The camera gets so close to one of these moments that the bear's hot breath fogs up the lens, a somewhat risky choice in the scene because it sort of breaks the spell by reminding you that this attack isn't "real," that it is being performed and filmed, probably computer generated; but it is also a craft choice that immerses you in the scene such that you feel the bear's breath on *your* face. You feel your own vision fog. You become a sentient camera lens. And you can't deny that you feel something for this bear—an intense level of emotion for this CGI creation—that seems inappropriate.

The bear is just trying to protect her children. At one point, she even pauses the attack to check on the cubs and make sure they're safe before resuming her mauling of Glass. I know the feeling. She's just doing what any parent would do. But she does it with such power, such strength and rage, and such commitment to savagery, part of me has to admire her. Part of me wants to *be* her.

She thrashes DiCaprio around as if he is a toy, ripping at his flesh, and pounding on his back with her paws, stomping him into the dirt. The two of them embrace again and again in this repeated act of consumption, and it seems impossible that either of them will survive to see their children.

When she's been shot and mortally wounded, the bear charges, reluctantly, almost out of obligation, and DiCaprio plunges his long knife into the bear's side, shoving the blade up into her heart or some other vital organ. Blood pours from her wounds, the two of them tumble down an embankment, and you feel a mixture of relief and sadness. It's over now. Or it's just beginning. And you feel your breath caught up in the top of your throat; you have to remind yourself to let it out. You have to convince yourself that it's safe to breathe again.

The whole attack doesn't last long, but it feels like forever. And I realized that, as I watched it in the theater, I'd pitched forward in my seat, my hands gripping the hand-rests like they were the safety bar on a roller coaster; and I made little noises, as if I was witnessing the attack live. When it was over, I turned to my friend and mouthed the words, "Holy shit."

It sounds crazy to say this, but not only did I want to watch the attack again and again (and I have since), but part of me wanted to experience it firsthand. The cinematography and sound editing, the acting and special effects, all made it feel so real, so immediate and visceral, I wanted to jump into the scene. I wanted to feel the bear's hot breath on *my* neck. I wanted to smell the deep woodsy stink of the bear's coarse fur and its hot blood spilling over me. I wanted that kind of intense ecstatic experience—which is not necessarily to say that I wanted to die or even be mortally wounded. I wasn't remotely suicidal. I just wanted to be close to the terror, to feel the energy of those precious moments. My friend just wanted the scene to end. At one point, she turned away from the screen, toward me, and I could offer her no solace. I couldn't break my focus.

I realize that these do not sound like the thoughts of a rational person. These should not be the thoughts of an overweight writer, a classroom volunteer, a professor and member of professional organizations who has bad knees and wears sweatpants a good part of every day. These are not thoughts I even entirely understand. But I have them and I cannot deny their existence; at least part of what I'm doing here is trying to normalize these thoughts and complicate the stories we tell about this kind of thinking and this urge to witness.

Sure, the attack scene in *The Revenant* is gruesome, savage, and terrifying—the sort of film scene that might make some people afraid to go into the woods in the same way that *Jaws* made people afraid of the ocean. In many ways the star of this particular scene is the bear, while DiCaprio gets the rest of the movie to shine. The scene is also strangely intimate. Personal, even. Seductive, mythical, and spiritual in its implications.

It's so visceral, so immediate and intense that it almost feels surreal. It is an impossibly artful creation of a bear attack that I will remember forever, catalogued into the archive of iconic, can't-forget movie scenes. Even though I know it's not true, I want to believe that this attack actually happened and that I honestly witnessed it; and at least part of this is because I believe the scene speaks to a very real and very human compulsion toward animal savagery. It speaks to the urge that many of us feel to have—or at least to witness—such ecstatic experiences. It's that urge, however taboo, to leap into an encounter with a force beyond our control, perhaps even beyond our comprehension.

It is perhaps not surprising, in response to this scene and the onscreen connection shared between DiCaprio and the bear, that Dwayne "The Rock" Johnson and comedian Kevin Hart rapped about it at the MTV Movie Awards. Surrounded by backup dancers in bear costumes, the cohosts rhymed about other 2016 movies, but returning each time to the chorus, "You'll always remember where you were . . . when LEO GOT FUCKED BY A BEAR!"

It's funny. Other minor celebrities stand up to join the fun, reciting their own memory of where they were when LEO GOT FUCKED BY A BEAR, and everyone laughs. It's quite a show. You can watch the whole thing on YouTube. But here's the thing: I believe the intimacy of this scene scares the average person more than the violence or gore, more than the undeniable terror of being attacked by a bear. They make weirdly homophobic jokes and perform this ridiculous rap with the backup dancers because, if they're being honest, The Rock and Kevin Hart and all the others also want to be "fucked" by a bear.

Okay, so what they secretly want—what many of us want—is perhaps less like being violently raped by an apex predator and more akin to the French concept of *jouissance*, which implies a kind of ecstatic experience, a mixture of pleasure and pain that shatters the self and, thus, provides an opportunity to reassemble oneself. It's kind of like being *fucked* existentially, emotionally, and intellectually, perhaps also physically. It's not a death drive, no Thanatos, or suicide ideation, but it *is* perhaps a drive to be destroyed or disassembled and then remade again. It's a desire to be fucked to death and to be reborn.

I think this is part of what David Villalobos wanted—or part of what I want to take from his story—and I think this destruction and rebuilding of the self is also at the heart of the audience's experience of the bear attack scene in *The Revenant*. To be clear, I'm not arguing that Hugh Glass felt pleasure as he was being brutally mauled by a grizzly bear; but I am suggesting that this is what we, as audience members, feel when we watch the attack—pleasure mixed with pain and repulsion. This ecstatic *jouissance* is what tingles through our bodies as we pause, rewind, and replay the scene over and over again. This *jouissance* is what frightened the writers of the rap into calling it "fucking," because there *is* something vaguely pornographic or at least voyeuristic about it; and you feel a little dirty for watching.

The bear, though behaving monstrously, does not necessarily come across as a monster, not in the same way that the shark did in *Jaws* or that some horror movie killer might scare us. She is just a bear being a bear, a mother protecting her cubs. She becomes both beast and phenomenon, both animal and annihilation. She is the hunted, not the hunter in this story; and the hunters are all white men, most of them weak, vile, or repulsive in some way. She is, in fact, one of the few female characters in the whole film. This bear is not a villain; that role is reserved for Tom Hardy's character, Fitzgerald. This bear didn't want or deserve this violence. This mother bear—this sublime and massive maternal creature—relied on savagery as protection. When your children are threatened, you do what you have to do. You don't start the fight, but you finish it. You fuck up some asshole who gets between you and your kids.

At the end of the scene, the sow lies there dead, her thick brown mass sprawled out on top of DiCaprio's mangled body, and her cubs are now left without a mother and a protector. You can hear them calling for her as the other men show up to pull the bear off of Glass. Their cries echo in the forest. The men pull her great mass off of Glass and she rolls over and flops onto her back, her head tilted down toward the camera. I can't help but feel sorry for the bear. I don't want her to die, but I know she has to for the sake of the movie. I know that it makes a better story if the monster dies and the hero survives. It makes the story a tragedy. But part of me wants the typical horror movie trope where she rises from the dead, lets out a monstrous roar, and savagely mauls three or four other men before finally dying at Glass's hand.

////

A WEEK OR SO before I was expected to play the role of an attack victim, the teacher sent me a copy of the original AP wire story that the students had read. He also sent me some follow-up details. I sat down and read through my script, trying to imagine what it must have been like, what this man must have seen and heard, my brain already working over the details and reaching for the unique subjectivity of the experience. I started doing some research and pretty quickly lost myself in story after story of bear attacks in the United States. I wrote pages of notes and obsessed over my character and all of his possibilities. I realized at one point that I was dreaming almost nightly of bear attacks.

I emailed the teacher and confessed my nervousness at "acting" for the first time, particularly at the challenge of embodying the subjective experience of an attack victim. I was worried over the weight of responsibility. But he did his best to reassure and prepare me.

"Don't worry," he said. "This will be fun."

PART ONE

STEPHEN

HAAS

BEING STEPHEN

The mountains, the forest, and the sea, render men savage;
they develop the fierce, but yet do not destroy the human.

–Victor Hugo

MISSOULA, MONT. (AP) — One hiker is dead and another
hospitalized after a bear attack in a remote backcountry area of
Glacier National Park.

National Park Service Rangers on Sunday night found Stephen
Haas, 37, of Yakima, Wash., huddled in a cave and barely conscious, suffering from multiple injuries brought on by an attack by
a grizzly. Haas's hiking partner, Janey Craighead, 50, of Moses
Lake, Wash., was found dead in a brushy area several hundred
feet from the cave.

Haas was in serious but stable condition Monday at St. Patrick
Hospital in Missoula, according to a hospital spokesman.

The attack occurred near Florence Falls in the Logan Pass area of
the park, south of the Going-to-the-Sun Road.

Rick Acosta, a Glacier National Park spokesman, said that Haas and Craighead were on a three-day backpacking trip when a large grizzly attacked them early Sunday morning while they were asleep in their tent. Haas escaped the tent, but Craighead was trapped inside.

Haas threw rocks and sticks in an attempt to scare off the grizzly, which then pounced on him, according to Acosta. Haas received a concussion, a punctured lung, a sprained wrist, several broken ribs and numerous cuts and bruises in the attack, the Glacier National Park spokesman explained.

No additional details were immediately available.[1]

---

1. The above scenario is the initial one provided to the students. In advance of the interview they were also given the following additional details:

From St. Patrick Hospital:
- Craighead died of a ruptured aorta.
- Craighead was the divorced mother of one daughter, Brandi Craighead, 25, of Spokane, Wash.
- Craighead was a captain in the Moses Lake Fire Department.
- Haas is single. He sells radio advertising for KYAK-FM radio in Yakima, Wash.
- Haas may be released from the hospital in two or three days.

From Glacier National Park spokesman Rick Acosta:
- A backpacker on Sunday morning notified Rangers of a possible bear attack. A search party found a torn nylon tent in a clearing around 3 PM Sunday. Inside was the body of Craighead.
- Shortly after, searchers heard a faint cry coming from a small cave nearby. There they found Haas, bleeding but conscious. An evacuation helicopter transported Haas to St. Patrick Hospital in Missoula.
- "It's just pure luck that he was found," one searcher said.
- This is the first fatal grizzly attack in Glacier National Park since 1972. It is the second reported attack of 2005.
- Searchers have not located the grizzly.

I knew I'd be asked questions about the attack, and I understood that the reporters would be looking for updates or clarifications, angles that weren't in the original story.

The instructor had made it clear that, as Stephen Haas, I didn't have to be an easy interview, and that I could try to derail their questions or be evasive. It was their job to ask the right kinds of questions, to steer my comments into something they could use. He told me that in a previous year the man who'd played my role had gone to the lengths of wrapping his head in gauze and bandaging an arm in order to look the part. This guy was the Leo DiCaprio of class visits.

I made no real effort at looking the part. I wore a baseball cap pulled down low over my eyes, and I carried a cup of coffee. Though I'd never really had an intimate or violent encounter with a bear, a costume seemed less important than being able to capture the emotional and intellectual reality of what it's like to survive a bear attack. I thought I should first call upon some of the things I knew about grizzly bears and Glacier National Park.

I also wrote out an entire life story for Stephen Haas . . . because that seemed necessary. And because I tend to over-prepare for anything like this, or because I'm more "method actor" than I care to admit. It's true that I felt as if I needed to get close to the subjectivity of Stephen Haas's experience, even if only through my imagination. I needed to become Stephen Haas.

I figured that it would help my story that—in real life—I had actually stopped in Glacier for a couple of nights on my way to Alaska in 1995 with my girlfriend (the girlfriend who would later become my wife and the mother of my children, and the woman I would eventually have to divorce after almost two decades of marriage and partnership). I thought my knowledge of the place would add some authenticity to my performance.

At Glacier National Park we'd stayed in a campground that had recently been invaded by a large grizzly. Signs had warned us to be vigilant about keeping food, or anything with an odor, outside our tent. It rained on us for two days and we day-hiked in a downpour, scared the whole time we might run into a wet and grumpy bear. We slept in the truck because it was easier than trying to set up a tent in the rain. We never saw any bears or signs of bears, but the traces of their presence seemed to float around everywhere like mist, like something you take into your lungs and it fills you up. And I figured I could use this kind of feeling in my performance. I could talk about the fear and the exhilaration tingling in my extremities, the way every sound seemed amplified.

FOR MY ROLE, I tried to imagine what Stephen Haas must have felt and thought at the time, tried to create him as a character in my mind. I wanted to tap into the subjectivity of the experience, which meant that I spent a lot of time reading reports of bear attacks, the wash of them sweeping over me until I could spit out facts and figures like Rain Man.

I also had my own questions for Stephen. I wondered how he'd managed to escape the tent when Janey didn't, and how he found the cave nearby. I wondered about the nature of their relationship. Janey was older and had a grown daughter. How did they know each other? Had they been camping together before? What made them want to be in grizzly country together? Did they discuss the danger? Was this trip a kind of test of their relationship?

Q: *Mr. Haas, do you have any previous experience hiking or backpacking in bear country?*

She trusted me, you know. I mean, I promised her that nothing would happen. I've been around bears before and they just . . . well, they don't normally care about humans. Or they don't so much in Alaska. I know it's different here, different in Glacier. I know the history of this place.

What history is that, Mr. Haas?

Attacks, man. Grizzly attacks. Night of the Grizzly. All that. I mean you can probably look them up. People don't get killed by grizzlies in Alaska. You know that, right? I mean you know that more people are killed by moose every year? That's because the bears and humans have adapted to cohabitate, to survive and thrive together. I mean, there are attacks. Sure. But not like in Glacier.

What is it about Glacier that makes it different?

It's a darkness. Something in the history of this place. Like in the DNA. There's been a tension between bears and humans here for a long time. Maybe always, certainly ever since that night those two women were killed. You know about that, right?

NIGHT OF THE GRIZZLY

Maybe part of the story of Stephen Haas and Janey Craighead, or at least my reading of it, began on a summer night, August 13, 1967, in Glacier National Park, Montana, four years before I was born.

On that night, two nineteen-year-old campers, Julie Helgeson and Michelle Koons, sleeping ten miles apart from each other, were each attacked and killed by two different grizzly bears; and their stories would forever change the way many people thought about one of America's last great predators.

MY FATHER USED TO tell me quaint anecdotes about visiting Yellowstone Park and of feeding curious bears from the car or watching the beasts rummage through a garbage dump. Bears were treated as docile mascots and encounters with humans were not only allowed but even encouraged. Bears were harmless, slow-footed cartoon characters, fun to watch as they pawed through the Yellowstone garbage dumps or ripped into your neighbor's cooler; they were the Park's resident entertainment, Yogi Bear and Boo-Boo, postcard characters of an endangered species. They even fought forest fires.

Yellowstone and Glacier parks are two of the only protected areas in the lower forty-eight states where grizzly bears still exist, but there are no more garbage dump feeding-times to watch, no drive-up encounters with bears, and visitors are regularly reminded that bears are dangerous and unpredictable and that, because they've chosen to drive, park, hike, or camp in the bear's home territory, they are assuming a certain amount of mortal risk.

That night in 1967 in the backcountry of Glacier, a grizzly stalked Julie Helgeson and her friends throughout the dark hours of the night, circling the camp, hunting them. They could hear it pacing, huffing and sniffing, and watching. They'd tried to keep it at bay with fire and a makeshift fort of logs, but the bear wouldn't leave. They heard it stomping around the perimeter of their camp, waiting for an opportunity. All night long, they kept their vigil, hoping the bear would lose interest in them and leave. But the bear didn't lose interest. The bear waited. And waited. And meanwhile, ten miles away, in the middle of Julie Helgeson's long night, Michelle Koons was dragged from her tent and killed by a different grizzly bear; and as a result, this night would never die, would always burn in the popular imagination.

It was speculated afterward by some that the bears had been attracted to the women's menstruation, as if grizzly bears were land-sharks, magnetically pulled to a single drop of blood; and while it is true that bears and other predators can smell blood from great distances, it was highly unlikely that this is what led those bears to kill that night. The story expanded, first in news reports of the attacks and then through a sensational and popular book by noted true-crime author Jack Olsen published two years after, titled *Night of the Grizzlies*, and later in subsequent documentaries and films that borrowed the title (or variations of it) and certain details of the events that night.

Most recently a 2010 documentary, titled *Glacier Park's Night of the Grizzlies,* revisited the attacks and interviewed survivors and park rangers who'd been there, making connections between the events of that night and the fate of the grizzly bear in the American imagination. In a sense, everything changed that night. Perceptions changed and policies changed. The story inflated into a myth that persisted for decades, floating in the margins of other attack stories, lingering also perhaps as a reminder that sexism perpetuates everything, even animal attack narratives.

Hegelson and Koons were, in essence, "blamed" for their attacks. Or at least that seemed to be what the stories were saying; and I wondered, as I prepared to play Stephen Haas, if anyone would ask me a question about Janey, about whether she was menstruating and if I thought that might have attracted the bear.

Part of me wanted to bait them a little, make obscure references to Night of the Grizzly, and see if the cub reporters would bite. Part of me wanted to know who'd done their research and who hadn't. If they had, they'd know about *Night of the Grizzlies*. They'd know that these were the first fatal attacks since the park had opened in 1910. And they'd know that since those attacks, Glacier National Park has been the site of other attacks, some of them fairly recent. They'd know that, statistically, your chances of encountering a grizzly bear in Glacier were perhaps greater than they were in any other National Park. They'd know you don't go to Glacier without knowing the bears are out there.

Q: *Mr. Haas, if you knew there were bears in Glacier Park and knew the history of attacks, why would you choose to go there?*

Have you seen the beauty of Glacier? It's a spiritual place, man. You should go. I mean, there are dangers everywhere.

Sure, but aren't you increasing the probability that you might be attacked by a grizzly bear when you hike or back-pack in Glacier National Park?

I suppose so. But you're also increasing the probability that you will experience something amazing, something so far from what you know in everyday life. It can reorder the way you think. It changes you. I guess that's a risk I'm willing to take.

Again?

Sure. I mean, not like anytime soon. . . . Do you know, have they found the bear?

BEAR ORIGINS

To name something is to own it, to shape its identity through language and to control it. The grizzly bear's scientific name is *Ursus arctos horribilis*, or the "horrible bear." Even its origin is monstrous and mysterious, the beast's very existence associated with terror.

When we describe a particularly gory accident or scene of violence, perhaps the death of someone who jumped into a tiger's or polar bear's cage, or even the scene of Janey Craighead's death, we might describe the scene as "grisly," a homophone that, at least for me, always conjures up an image of a "grizzly" bear. The word "grizzly" may descend from both "grizzled," meaning "gray," or old, as well as from the Old English, *grislic*, meaning

> . . . *"horrible, dreadful" from root of grisan "to shudder, fear," with cognates in Old Frisian grislik "horrible," Middle Dutch grisen "to shudder," Dutch griezelen, German grausen "to shudder, fear," Old High German grisenlik "horrible," of unknown origin; Watkins connects it with the PIE root *ghrei- "to rub," on notion of "to grate on the mind."*

I try it in a sentence: *Stephen Haas shuddered with fear before the horrible, dreadful bear and the grisly scene that awaited him the next morning.*

I BEGIN TO UNDERSTAND the way language might shape this moment, and perhaps how it continues to shape the second-day stories of this incident and others like it. But it is this last root meaning, this connection from Calvert Watkins, editor of *The American Heritage Dictionary of Indo-European Roots*, that has planted itself deep within me: this idea that the animal and its name, the sign and signifier, each mean to grate, to grind, and worry the mind. A scene, an image, a story is grisly and *grizzly* because it lingers, because it takes over your thoughts at times, lurking with others in the recesses of your consciousness. It is grisly because it never leaves, because it haunts you, attacking the quiet moments.

I REMEMBER AS A kid hearing about the attacks in Glacier National Park and, though I couldn't find it in my research, I know I've seen a clip from some horror-movie adaptation wherein a monstrous rampaging grizzly attacks a woman in a mummy-style sleeping bag. He picks her up as if she is light as a leaf and throws her against a nearby tree. I can still remember the images, the scene playing out in my mind. And though this was something of a special-effects exaggeration designed to make the bear seem even more monstrous, it wasn't *much* of an exaggeration.

After finally charging the camp in the early morning light, the bear dragged Julie Helgeson in her sleeping bag out into the brush, where she was quickly mauled and killed as her campmates listened.

The grizzly bear was that dark malevolent force—unseen, unpredictable, and unstoppable; and one of the last great predators in North America, one of the few species that we hadn't yet hunted to extinction. Though bears and mountain lions have now made a comeback in many parts of the country, most of that hadn't begun yet when I was growing up in the '70s and '80s.

The Glacier Park bear's behavior troubled park rangers and scientists because they hadn't ever heard an account wherein the bear clearly seemed to be stalking and hunting humans. Bears weren't supposed to act this way. They didn't hunt people. Or at least they weren't supposed to hunt people, not according to previous research. But both bears that night exhibited unusually aggressive behavior that was difficult to reconcile with what humans thought they knew about grizzlies in the park, proving ultimately how little we actually knew about the predators around us.

Native American cultures have long acknowledged the powerful symbolism of the bear, recognizing its wisdom, courage, power, and strength, as well as its unpredictability. Some cultures also see the bear as a healing maternal spirit with powerful medicine, a peaceful solitary mother figure who is, nevertheless, capable of great savagery and aggression when provoked or threatened. The bear is never just a bear. It is always something greater and more wild, more sublime and powerful than humans can perhaps ever fully understand.

Q: *Mr. Haas, can you describe what you remember of the attack?*

I don't sleep very well most nights. I slip in and out. I hear things, you know. I always thought I'd hear a bear coming, that it wouldn't be like a shark attack that you don't see or hear, that it would be like a tornado and sound like a freight train coming . . . but I didn't hear anything until the bear was on top of us. It's hard to describe . . . the tent fabric ripping and the fiberglass poles snapping. You don't really know what's happening. But you can smell the bear.

What did it smell like, Mr. Haas?

Like wet earth.

The earth?

I got tangled up in the tent fabric, trying to get out. I fell. I remember hearing Janey scream . . . it was horrible. I don't know what happened after that. I tried to go after her. I'm sure I did.

Was that when you hit your head?

My what? Oh, right. My head. I guess so. Like I said, I don't remember much after that.

Were you injured by the bear?

I don't know. I don't remember how I hit my head. I want to believe that I was trying to fight the bear off, trying to get it away from Janey, you know. That would make a better story for you. For her daughter. I wish I could tell Brandi more. I wish I knew more. The doctors said it looked like I'd hit my head on a rock.

Why do you think the bear went after Janey instead of you?
It would be good if I said that I wish it had been me. But that
would be a lie. I mean, how many of you would volunteer to
be attacked by a bear? Who does that sort of thing? I wish Janey
hadn't been attacked, wish she hadn't died, and I wish that bear
didn't have to die . . . but I wouldn't trade places. And neither
would you. I'm just being honest.

BEAR CONFESSION

There's another truth in this: I'm somewhat bear-like, ursine in my personality and presence. I am an apex predator, even if I try not to act like one most of the time. I can be the roaming bear cruising the tundra, grazing and fishing, but not the tiger stalking, the cat pouncing, or the wolf circling. I do not hide well, and I'm not interested in bullish aggression for the sake of aggression. My size is my shield and my show, the protective bubble you do not want to break. Like a bear, you'd have to provoke me to fight. But I don't really know what I might do or of what I'm capable in such situations. Sometimes this frightens me. Sometimes I'm afraid I will be tested.

At 6'4" tall and weighing in at around 260 pounds, I sport a torso like a beer keg and stout legs that are too short for my height; my ankles and wrists are surprisingly thin—probably a gift from my mother, but I wear XXL shirts and have close to an eighteen-inch neck. I can't really grow much facial hair, but I have a scar on my right cheek and, to some people, I look big and intimidating, especially when I buzz my hair down to stubble.

In bars or at parties in high school or college, when the alcohol was flowing, I was often mistaken for the alpha male type who wanted to play-fight or wrestle like juvenile bears at a salmon stream. But I don't like to wrestle. I don't play-fight. I don't want to slap or punch and pretend that violence is an appropriate substitute for affection. Hugs are okay. But I don't even want to yell and scream when I'm drinking. I don't want to be jostled or touched or shoved too much, especially by strangers.

When I lived in a house full of guys in college, my roommates would often wrestle and roughhouse with each other. I asked one of them why they never included me in their games and he said, "Because you'll hurt us."

I don't know if this was true. It might have been. To me violence was not play. Part of this was undoubtedly due to the lessons I learned growing up big. When I did the same things other kids did, they had different consequences. When I roughhoused, I hurt people, even if I didn't mean to do it. And often because of my size I was a target, a test for some kids to see if they were brave enough to mess with the big kid.

Once, at a cousin's friend's house, an older boy decided he wanted to test me. He kept teasing me and hitting me, and eventually wrestled me to the ground and pinned me down, putting all of his weight on top of me. He was laughing at me, calling me fat. And I didn't like it. He'd gone too far. So I grabbed his legs and wrapped my arm over his neck, then I stood up, holding him up on my shoulders. And like something I learned from professional wrestling, I lifted him over my head and slammed him to the floor. After that he crawled under a bed and wouldn't come out for a while.

Unfortunately these are lessons my daughter is learning for herself now. At seven years old and well over four and a half feet tall, pushing five feet soon, she's taller than almost every other kid in her grade level and towers over the boys. I've told her what I learned growing up big, that she has to avoid physical confrontations because if she pushes a kid, even in a simple game on the playground, it can have different consequences. It can mean bruises and parents talking. She has to be like the bear, playful but solitary, self-possessed and rarely predatory.

But I've also told her what my father always told me:

"You don't ever start a fight. But you finish it."

Be the bear, not the bull.

Q: *What can you tell us about the cave where you were found, Mr. Haas?*

I don't remember the cave. Just the shadows. And sounds. It felt like I'd drank way too much whiskey. Everything was blurry and I kept blacking out. The noises were bad. How long was I in there?

All night, Mr. Haas. The rescuers found you in the morning. Some hikers had stumbled across your campsite and could tell something bad had happened. That cave, though, was pretty convenient, huh? Had you scouted it out before as a possible shelter?

Cave is a little generous, really. It was more like a hole between some bigger rocks. I might have hit my head diving into the rocks, just trying to get away. I remember waiting for the light to change. I wanted the sun. And I wanted to find Janey.

BASED ON A TRUE STORY

In 2007 there was a Stephen Haas who sold advertising for KYAK-FM radio in Yakima, Washington, which now appears to be a Christian radio station catering to a largely Latino population. This may or may not have been true at the time that Stephen Haas was attacked by a bear. This may or may not have even been the same Stephen Haas. In fact, the deeper I got into my preparation to play Stephen Haas, the less real he became.

I read again the AP news story that the teacher had given me, but I could find no record of a similar attack in any newspapers or other sources. Craighead's death was not listed on a fairly comprehensive list of fatal bear attacks in North America. In fact, in the course of my search, I could find no record for anyone named Janey Craighead at all. Moses Lake, where she was supposedly a captain for the fire department, is a small town in Central Washington, 377 miles from Glacier National Park, but only about 100 miles from Yakima, where Stephen Haas may or may not have lived.

There did seem to be a Rick Acosta who worked for the National Park Service but it wasn't clear if he ever served in Glacier National Park; and though initially I (perhaps foolishly) assumed the story to be true, I slowly came to realize that the whole thing was most likely fabricated or cobbled together from other stories for the purposes of the class. It was a fiction. A fake. A lie of sorts.

For some reason this disappointed me, and not because the teacher used fiction to teach students how to report facts in a journalism class. It disappointed me to know that I'd put all that work into pretending to be a fake person, a character in a fabricated attack story. I believed that I was pretending to be a real person. I wanted to be part of a "based on a true story" story because I like the way the truth bends to the details in those stories, and because these have always been my favorite kinds of stories; and I guess I wanted to feel that all of my research and writing was an effort at intimacy and empathy with a real person who'd suffered real loss.

I eventually realized that it was just my own subjectivity, my own imagination at work. Stephen Haas could only be as real as I made him. He was just a vessel. It was a lot of pressure caring for such a container. Suddenly I felt responsible for making my performance convincing, for creating a whole person out of the ether, a character with a parents and a hometown, a backstory that informs his present existence. I was the only one responsible for telling the story of Haas and Craighead and for keeping them alive.

This only partly explains why I was reading a lot of bear attack stories and eventually found myself thinking about the "Drama in Real Life" sections of *Reader's Digest* that I used to consume as a child during visits to my grandparents' house, many of them recounting dramatic and harrowing tales of animal attacks and narrow escapes. I loved these stories. You can still find them online and in the magazine. They pretty much always end the same way, with hope and survival against great odds. The stories are inevitably a kind of celebration of the indomitability of the human spirit, combined with a graphic portrayal of the savagery of a morally indifferent natural world. And bears. Lots of bears. I figured these "Drama in Real Life" stories were exactly the sort of story I needed to tell if I was going to be convincing as Stephen Haas. It seemed I should be able to answer my own questions, however uncomfortable. The problem, however, became knowing where to begin and where to end.

Q: *Mr. Haas, what do you think you learned from your encounter with the bear?*

Learned? About what?

Perhaps about yourself or about bears, or maybe about the safety of backpacking and camping in Glacier National Park?

The bear was just being a bear. We were in his territory. We'd leaped into his cage. And the point of such leaps, I suppose, is fundamentally selfish. I mean, it's about you, ultimately. It's about testing oneself. It's about being humbled.

What do you mean by cage?

Look, think about it this way. How do you define a cage? The only differences between a zoo and a national park are the size of the cage and the consequences of leaping into that cage. In a zoo, the probability of attack is increased exponentially, but the difference between a zoo cage and Glacier National Park is mostly a matter of percentages. You know what I mean?

I'm not sure I do. Are you suggesting that our national parks are essentially very large zoos where we are allowed to climb into the cage with the animals?

Yes. Yes I am.

ATTACK RESPONSE

Though it's not a pleasant thought, a bear will, more often than not, eat a person it has killed. Stephen Haas spent the night injured in a cave, hiding from the bear; but what did he hear, what did he hide from, and what did he face there in the dark?

The question I don't want to face: Did the bear eat Janey?

The question I would ask: Did the bear eat Janey?

To put oneself in proximity of bears is not to simply risk bodily harm but bodily consumption, a true kind of communion with the wild. I didn't know if the bear who'd killed her had eaten Janey, or part of Janey. I wasn't given that information. But I could imagine the scene. I knew how these stories often went and I was prepared to offer up some grisly details.

Unlike the case of Janey Craighead and Stephen Haas, most bear attacks occur on trails, when a hiker or backpacker surprises a sow grizzly with cubs. It's even more rare for a bear to attack people in a tent. When a bear does attack, it will often go for the head or neck. A great many people who survive a bear attack report having their scalps ripped off and their skull chewed upon. Often the bear will jump up and down on top of a person, breaking ribs and knocking the wind out of him. Often the bear will break your neck. Sometimes, if you play dead, a grizzly will leave you alone. Sometimes they will kill you and eat you no matter what you try to do.

The hard thing to accept is that most of the time a bear is not hunting for humans, not even killing because he thinks he can eat the human. That comes afterward. Usually a grizzly attacks because it is protecting its young or guarding a recent kill—often a moose calf or caribou or some other ungulate it has run down; grizzlies are rarely predatory toward humans. They simply don't often see us as food. We're too much work. Too noisy and weird. We're all limbs and skin— like brightly colored flightless birds awkwardly shuffling through the brush. We're hardly worth the effort, so little meat compared to other animals; and besides all that, we often try to shoot them, spray them with pepper spray, or otherwise antagonize them.

Fatal bear attacks are actually extremely rare; but unfortunately such attacks are more common in the lower forty-eight states than in places like Canada and Alaska, mainly because there are more peo- ple, and bears have, in some places, learned to associate humans with food. When attacks do occur, bears that kill and partially eat people are typically then killed, sometimes because they hang around the area, protecting their food, sometimes because they can't get away fast enough.

These bears are killed because they've now developed a taste for human flesh and see humans as a food source and could, thus, become predacious; and if an apex predator who lives in proximity to humans decides to hunt them, there is little that can be done to stop it, aside from killing that predator.

Survival is a wild bear's only imperative; and they're mostly unconcerned with our reasons for being in their proximity. Bears who attack humans are sometimes killed to confirm identity, necropsied and examined for evidence; and they are killed for revenge, killed because a bear can't often kill a human without paying for it, even if a bear is just doing what bears do—operating on instinct, or even if it's just a bear unlucky enough to be in the same area. One attack at

Glacier National Park prompted the killing of five bears before it was believed they finally shot the one responsible.

I thought I might ask the student reporters if they knew this. I thought I might ask if they knew whether the bear that killed Janey had been found, and if it had been killed. Sometimes they have to open the bear up and look inside his stomach to be sure they got the right one.

"There's the answer to your questions," I'd say. "You have to look inside the bear."

Q: *Mr. Haas, can you describe what you saw when you emerged from the cave?*

Shadows mostly. The light seemed like it had been sucked out of the sky. I heard voices.

The hikers who found you?

Sure. The hikers.

Anything else that you remember?

Have you ever seen those Buddhist prayer flags? You know the colorful ones hanging from a rope? The tent looked like that. It was kind of spread out across the alder, all torn up.

Were you able to locate Ms. Craighead?

I like that movie, Apocalypse Now. *You know the scene I'm talking about. That one with Brando and Sheen where Brando is telling him, "You must make a friend of the horror." That's such a great scene . . .*

Mr. Haas?

I can't do it. I can't make a friend of that horror, not those images. I can't give them to you. Not yet.

MY SIDE OF THE MOUNTAIN

I realized at some point during preparation for my role as Steven Haas that Janey Craighead, my camping partner and the sole fatal victim of the bear attack, also happened to share a name with Jean Craighead George, the author of the 1959 young adult novel, *My Side of the Mountain*, a book largely responsible—in a somewhat indirect way—for my own trip to Glacier National Park and Denali National Park in Alaska in 1995 and for my preoccupation with a life of solitude in the wilderness.

At the age of eleven or seventeen, if you'd asked me what I wanted to be when I grew up, I would've probably answered, "River raft guide or maybe a hermit in the woods."

It was Craighead's story of Sam Gribley, who chooses to leave his large family and society in search of a solitary life in the woods that probably ignited much of my fascination with a solitary life in the woods. Sam carves a home in a hollowed-out tree and keeps a peregrine falcon named "Frightful," and this seemed like a pretty good plan to me.

Hovel in the woods? Check.

Animal for a best friend? Check.

I believe it may be books like Craighead's or Walt Morey's books, and later work from Edward Abbey and Jon Krakauer that inspire people like Christopher McCandless to actualize the fantasy, to break the chains of family, society, and home, and to leap into the wild; and I can't deny that I always found these stories undeniably compelling.

I don't mean to blame these books or authors, but rather to suggest that what makes them powerful and enduring works of art is that they tap into a deep and elemental desire of many people, especially it seems of young men in their twenties and thirties—that prototypical western drive to test oneself against nature, as if there is something hormonal or physiological about this drive, as if it's in our marrow and coursing through our DNA. Hell, I had *plans*. And I shared these plans with anyone who would listen, including my girlfriend.

Craighead's book (and Morey's *Gentle Ben*) certainly planted the seed for me, and Jon Krakauer probably nurtured it. My writing mentor in graduate school told me that nearly every semester he teaches Krakauer's *Into the Wild*, the perhaps largely imagined and embellished story of the life and death of Christopher McCandless, one or more young men in the class experience a kind of breakdown, sometimes even disappearing from class and school altogether, stepping away from life much as McCandless did and run for the hills. He tells me he has to be careful with that book.

He tells me that it is a dangerous book.

I love this idea. Because I think I've felt that urge, too. And that book *is* dangerous—a hypnotic essaying of this urge that so many young men seem to feel, a book that is both warning and siren call to chase the wild.

I'd already read Krakauer's 1993 piece on McCandless from *Outside* magazine, "Death of an Innocent," when, in 1995, after my graduation from college with a degree in philosophy, my girlfriend and I drove from Kansas to Denali National Park in Alaska. I knew we were going where the bears are as big as Volkswagen Beetles. I knew McCandless had died not far from Denali, but I didn't really know where. I knew also that part of Krakauer's mission in the article, and later in the book, was to normalize this urge to disappear into the wilderness and to rescue McCandless from a simplistic understanding of his death.

That first night, after we set up camp in a stiff wind, I pulled out my map and studied it for a while in the tent. It took me a while to see the full picture, but I eventually realized that we were camped on a ridge above the valley where McCandless lived and died in an abandoned bus. I hadn't planned this. *Or had I?* When we picked up our backcountry permit, we'd taken one of the only quadrants left available that had any elevation. We'd just climbed to the top of the ridge and picked a spot to pitch our tent.

The next day we hiked up and peered down into a vast stretch of trees and green bogs, and I saw the river that, swollen with runoff, had supposedly prevented McCandless from making it out when he'd finally realized he needed help. On the map I saw the hand-crank ferry that could have saved him if he'd only known about it, if only he'd walked a few hundred yards farther upstream. At some point between the looking glass and the map, I realized I was staring into a future book, a kind of alternate reality, and perhaps into my own possible future.

Between Krakauer and Craighead, I found literary inspiration for my own engagement with the wild, my own desire to lose myself in bear country. What I didn't realize some twelve years later as I was preparing to pretend to be a bear-attack survivor was that Jean Craighead was the younger sister of pioneering grizzly bear researchers, conservationists, and twin brothers, Frank and John Craighead.

Like Sam Gribley and his falcon, her brothers had caught and trained a Cooper's hawk. Their work with the bird eventually led to them developing a relationship with *National Geographic* and a series of TV specials that introduced America to the outdoors-loving, naturalist-nurturing Craighead family. The brothers pioneered the use of radio collars and tranquilizer darts to track, capture, and study grizzly bear populations in Yellowstone National Park. One of their great successes and unique talents was combining conservation with entertainment, particularly with documentary film. They were something like the Jane Goodall of grizzly bears, exposing whole generations of Americans to one of the last great predators of North America, and perhaps staving off all-out extermination of them in the lower forty-eight states.

Q: *Mr. Haas, were you aware that there were grizzlies in the area? Did you feel that you had ample warning and were adequately prepared for encountering bears?*

Look, we went there to encounter bears. You don't go to Glacier if you don't think that might be a possibility. We wanted to see bears.

Did you and Ms. Craighead see warnings from the national park about bear activity?

Warnings? Like signs or something? The whole place is a warning. The trees whisper of it. You can't be there without feeling the presence of bears. You can't stand on a lightning rod and feel surprised when you get struck.

Why would anyone stand on a lightning rod?

Humility. How many times in your life have you sought out an experience that humbles you? You know what I mean? How many times have you, sir, ever been in a place where you are not the top of the food chain? Do you know what it's like to be prey? It's strange, really. And pretty cool. Kind of hard to explain. I guess it's a unique intimate experience, a chance to know yourself better.

HOW TO SURVIVE A BEAR ATTACK

I wanted to toy with the student reporters a little, to lead them through tangents and digressions, to try and get their second-day stories off-track. I figured I'd just mention Alaska a lot, knowing full well that the attacks occurred in Montana. But what I also really wanted to tell them is that in Alaska, where the lines between civilization and wilderness are blurry at best, and nonexistent at worst, more people are killed by moose most years than by grizzly bears or black bears.

Many of these moose-related deaths occur in yards, neighborhoods, parking lots, and along the roadside. Moose often graze along the roadways, where the weeds are kept back, the trees trimmed, and lush green grass is plentiful; and quite a few of these deaths have occurred because a tourist stopped his car to get up close to a moose or a moose calf for a photograph and was stomped to death by a 1500-pound animal that kicks with its hoof-armored front legs.

"Moose are not horses with antlers," I would say. "They are not domesticated or docile. They will kill you. But they won't eat you afterward, which is a plus."

I thought that might get some laughs.

Then I figured I'd take them even further down this tangent and offer some advice if you should encounter an angry moose in the wild. They are a lot easier to escape than a bear. Several guidebooks I read offer tips for escaping a charging moose, but it seemed the best survival strategy was to put a tree between you and the moose. When it charges you simply run around the tree.

This sounded cartoonishly simple to me until I thought more about it. Moose can't corner. It's like a school bus trying to chase a motorcycle around a telephone pole. If only avoiding a bear attack were so simple.

ACCORDING TO A NUMBER of sources I read before and during my time in Alaska, the number one tip for avoiding bear attacks is to avoid looking or smelling like food. You're not going to outrun a bear; despite how fat and slow they look, a grizzly can run up to forty miles per hour. And you're not going to fight off a grizzly or overpower it.

It seems so simple.

Don't be food.[2]

But surviving a bear encounter is more complicated than you might think. I read a lot of travel guides for Alaska when we visited there, all of which made special mention of how to avoid attacks. The more specific following tips on how to survive, taken from a website (unironically and unfortunately) called *The Art of Manliness*, are fairly typical of what I found.

> **1. Carry bear pepper spray.** *Experts recommend that hikers in*
> *bear country carry with them bear pepper spray. UDAP bear*
> *pepper spray is a highly concentrated capsaicin spray that creates*
> *a large cloud. This stuff will usually stop a bear in its tracks.*[3]

---

2. Because the best way, apparently, to avoid acting like food was to make a lot of strange noises, many places in Alaska sold "bear bells," which were just cheap metal bells that you were supposed to clip to your pack or belt. Reviews seemed to be mixed regarding these bells with some anecdotal evidence suggesting that they simply sound to a bear like birds or a dinner bell.

3. Maybe. Other stories I heard in Alaska mentioned the likelihood that pepper spray might just aggravate a charging grizzly or, worse, blow back in your face. Most of the locals we encountered on the trails packed a different kind of heat—typically a .45-caliber Magnum pistol or a shotgun with lead-slug shells—but there was something about this that I found extremely unnerving. I didn't keep a gun in my house, and I didn't pack a gun when I was out in public, and I didn't understand why that should change when I was backpacking in the middle of the Alaskan wilderness.

2. **Don't run.** *When you run, the bear thinks you're prey and will continue chasing you, so stand your ground. And don't think you can outrun a bear. Bears are fast. They can reach speeds of thirty to forty miles per hour. Unless you're an Olympic sprinter, don't bother running.*[4]

3. **Drop to the ground in the fetal position and cover the back of your neck with your hands.** *If you don't have pepper spray or the bear continues to charge even after the spray, this is your next best defense. Hit the ground immediately and curl into the fetal position.*

4. **Play dead.** *Grizzlies will stop attacking when they feel there's no longer a threat. If they think you're dead, they won't think you're threatening. Once the bear is done tossing you around and leaves, continue to play dead. Grizzlies are known for waiting around to see if their victim will get back up.*

Surviving a grizzly attack is difficult but not impossible. It's unwise to surprise a grizzly, especially if he's eating or resting in the alder thickets; and cubs, as cute as they are, will never be too far from their overly protective mothers. Just ask Hugh Glass about this.

Within the first two weeks we were in Alaska, a grizzly attacked a local family of hikers outside of Anchorage. It was a well-traveled trail, one the family had traveled many times before and knew well; but a male grizzly had killed an elk calf near the trail, the sort of thing you can't really predict in Alaska, and the bear was protecting his cache of food. Anyone on that trail would've been a threat to his food.

Before he was done, the bear had killed a mother and her son and chased the woman's fourteen-year-old grandson up a tree. The bear

---

4. If you are, however, an Olympic sprinter, then, by all means, try to outrun a bear.

dragged off their bodies and ate part of the mother and her son before rescuers arrived and park rangers eventually killed the animal. The boy waited out the attack and listened; and I remember reading the story and thinking about him, about how the worst part must have been afterward—the waiting up in that tree and listening, the hour or more before help came, when he was alone with the bear and the bodies of his grandmother and his uncle.

Q: *Have you spoken with Ms. Craighead's daughter?*

I don't think so. My thoughts are a little jumbled. I don't know if Brandi wants to talk to me, honestly. There's no way I'm the one who survives. Janey was so much better in the woods than me, so much more comfortable.

How did Brandi feel about her mother backpacking in bear country?

What kind of question is that? I mean, how am I supposed to know that sort of thing? Do you imagine that there was a fight or something? Maybe Brandi hated that her mom spent so much time with me. Maybe she was jealous. Maybe she'd called her mom the day before we left and told her it was crazy for her to go backpacking in Glacier. Maybe she mentioned bears, fucking huge grizzly bears that can kill you. Maybe she mentioned her long-dead father and asked, "What if something happens to you?"

But did she say those things?

I'm tired. My head hurts. You don't . . .

Do you think you'll go back into bear country, Mr. Haas?
Will you return to Glacier Park?

I can't stay away. Perhaps I need some help. Perhaps you should come with me. That's where you'll find the second-day story. It's still out there, floating around. The ground is stained with it. The bears are innocent. They could never be anything other than innocent.

Have you forgiven the bear that killed Janey?

Of course.

BEAR AWARENESS

It would be ridiculous to say that I had gone to Glacier and Alaska because I wanted to be *attacked* by a grizzly bear. It would not, however, be an exaggeration to say that I wanted to encounter a grizzly bear in the wild, perhaps even in close proximity; and in preparing for my role many years later, I thought this admission might be something that could get me closer to Stephen Haas. You don't go to Glacier or Alaska, and you don't backpack there without knowing there's a chance you'll see a bear. In fact, many people travel there for precisely that experience of encountering a bear in the wild. They seek it out.

In Alaska I sometimes tried to reassure my girlfriend that being killed and eaten by a bear was a noble death, or that it would be a sign that it was probably our time to go. I told her that we'd be famous, that people would read about us in the papers. Maybe they'd even make a movie. Maybe actors would be hired to play us in the movie. She didn't find this line of argument particularly convincing.

My desire to come face-to-face with a grizzly began to get more complicated once I was actually living, camping, and trying to survive in bear country. It wasn't just that I'd brought someone else along with me. It was that the reality of encountering an apex predator in the wild moved from the realm of possibility and imagination into the realm of reality. And that reality began to feel pretty heavy after a while.

The visitor center of Denali National Park where we picked up our backcountry camping permits displayed numerous books and videos about bears and bear safety. One video playing on the monitor showed a massive grizzly the size of a small car running full speed across tundra in pursuit of an elk calf. The video said he was moving between thirty and forty miles per hour. I watched the bear's muscles rippling beneath his fur, his bullish haunches and beer-keg-sized head take down the helpless mewing elk calf like a brown wave washing over a seashell, and my guts quivered. My knees wobbled and I felt a little sick.

Sometimes when I'm driving my car at forty miles per hour, I'll still think about that bear. But more to the point, a day later as my girlfriend and I trudged up a hillside across tundra, an experience not unlike walking through a field of two-foot-thick wet sponges, I would remember the image of that bear and the ease with which he moved, and I would think more about what it would be like to actually encounter an Alaskan grizzly bear out there on the tundra, exposed and vulnerable.

As we labored up that hillside to our first camping spot, stumbling beneath the weight of our packs, I looked back at my girlfriend a few paces behind me. Half my size, the going was harder for her, and I remembered that bear galloping full speed across the tundra, and the half-joking advice my father had given me before we left on the trip: *You don't have to be faster than the bear, just faster than the slowest person on the trail.*

Q: *Do you have a family that is worried about you,*
Mr. Haas?

I have some family.

I'm sorry, Mr. Haas. Can you clarify that statement?

Two children. They live with me half the time. They're with
their mother now. I talked to them last night on the phone.
My daughter . . . she was crying.

I'm sorry, Mr. Haas. But can you tell us what she asked you
about what happened?

She asked me if the bear got our food.

Your food?

It's a long story. We went camping once. There was a bear.

A bear?

I don't want to talk about this right now. She was upset. That's
enough for your story. My son, too. Maybe even more so. It's hard
for them to understand. And they don't know the whole story.
They don't know about Janey. Or not everything. I didn't know
how to tell them.

BEAR CONFESSION

Here is what I wouldn't admit to the student reporters: We never did encounter a grizzly in the wilds of Montana or Alaska. We saw plenty of grizzlies from the backpacker's bus that we rode into Denali, and spotted other bears along the side of a road that snaked around a lake in the Yukon Territory. But the only personal experience I could really draw upon to get close to what Stephen Haas witnessed during his encounter was a brief run-in with a black bear, not a grizzly.

We'd been backpacking over Eagle Pass, just outside of Anchorage, and seen plenty of evidence that grizzlies were around. There were spots along the trail at the higher elevations where grizzlies thrived, and it was clear bears had bedded down in the brush, perhaps just taking a nap and waiting for their next meal to wander along. We found their tracks everywhere, along with piles of berry-laden scat.

Obeying the guidebooks, each time I came across a pile of poo, I bent down and put my hand close to the pile, feeling if there was still heat radiating off of it. If the scat was warm and steamy, that meant the bear wasn't far away. That meant the heat from the bear's intestines still lingered long enough for me to feel it fog my palm.

The trail wound through alder thickets that were tall and dense enough that you couldn't see around the next bend in the trail, and we kept seeing more and more grizzly tracks. The hike seemed to take forever; and while I put my hand close to a lot of poop, we never saw a bear.

We eventually made it through, but not without some tension and creeping fear, some real worry about what might happen if we did encounter a grizzly. Burdened with our packs, there was no way we could run or even move quickly. And the next day, as we descended into the valley, we started seeing black bears. Lots of them. We were trudging along a trail that followed a bend in the Eagle River, when one bear who must have weighed 200–300 pounds swam across the river, climbed the bank, and started running down the trail toward us.

I knew what the guidebooks said. I knew that black bears, even more so than grizzlies, had been known to prey on humans. I knew that you weren't supposed to run because it triggered their chase instinct. And I knew that if confronted by a black bear, you were supposed to fight back. The bear kept coming and we had a choice to make.

My girlfriend tugged at my sleeve, "Seriously. Let's go!"

I paused for a moment.

And then we ran.

I'll admit it here: We turned around and hightailed it out of there. We ran as fast as we could back the way we'd come, making our way to a rocky sandbar. Then we just stood there waiting for the bear to come bursting out of the trees after us; but it never did. We weren't sure what we were supposed to do next. We had no choice but to stay on the trail. To go back would mean we'd have to climb the pass again and spend a night in grizzly territory.

The whole thing was pretty scary. There's no doubt about that. I was sure we were moments away from some kind of really uncomfortable bear encounter. I'm not sure I ever imagined the possibility that one of us might be hurt. But still, I did not run *toward* that bear. I ran away, my heart racing and pounding in my ears. Afterward my senses tingled and every sound was magnified. I could understand why some people find the experience addictive. But I could also understand why my girlfriend wanted no part of it.

Soon a solo backpacker came up from behind and passed us on the trail.

"Careful," I said. "There's a bear up there."

He just smiled, patted the .44 Magnum hand-cannon he had strapped to his side, thanked me, and kept on walking.

After a while, when we didn't hear any gunshots or cries for help, we figured it was safe and kept moving. We never did see the bear again, but there were signs of bear activity everywhere. The trunks of trees along the trail were striped with claw marks and the trail dotted with piles of scat. There was something profoundly strange about being in *their* territory. I felt like a trespasser, an interloper, and an alien in someone else's world. But it was also thrilling in a way that's hard to explain.

My girlfriend and I barely spoke as we hiked the last few miles, ours senses attuned to every noise, every movement, every bear track or pile of scat. The quiet vigilance and hyperawareness required to hike in bear country was something I hadn't experienced before; and perhaps this feeling is part of what Stephen Haas was seeking out there in Glacier Park. Perhaps this was also close to what David Villalobos wanted—those oddly exciting and profoundly humbling moments where we're forced, as humans and predators in our own right, to acknowledge our lower status in the food chain but also our place in the wider world of the beasts who will inevitably consume us.

PART TWO

TIMOTHY TREADWELL

THE GRIZZLY MAN AND ECSTATIC TRUTHS OF MEDIATED WITNESS

I found that beyond the wildlife film, in his material lay dormant a story of astonishing beauty and depth. I discovered a film of human ecstasies and darkest inner turmoil. As if there was a desire in him to leave the confinements of his humanness and bond with the bears, Treadwell reached out, seeking a primordial encounter. But in doing so, he crossed an invisible borderline.

—WERNER HERZOG,
FROM THE VOICEOVER NARRATION
FOR *GRIZZLY MAN*

I'm out in the prime cut of the big green. Behind me is Ed and Rowdy, members of an up-and-coming sub-adult gang. They're challenging everything, including me. Goes with the territory. If I show weakness, if I retreat, I may be hurt, I may be killed. I must hold my own if I'm gonna stay within this land. For once there is weakness, they will exploit it, they will take me out, they will decapitate me, they will chop me into bits and pieces. I'm dead. But so far, I persevere. Persevere. Most times I'm a kind warrior out here. Most times, I am gentle, I am like a flower, I'm like . . . I'm like a fly on the wall, observing, noncommittal, noninvasive in any way.

Timothy Treadwell had become one with the bears. Or at least he said as much. And at times in Werner Herzog's 2007 documentary, *Grizzly Man*, it sure looks like Treadwell is an accepted—or at least tolerated—member of the grizzly bear community he embedded himself within. At other times, it's clear that Treadwell is playing a role, and that he is acting like a bear and internalizing their rules of intimacy and savagery; and he has crafted a powerful narrative at the center of which is Treadwell himself. His movies are an epic memoir, a thirteen-chapter, thirteen-year construction—and ultimately, destruction—of the self.

In the monologue above, Treadwell describes himself at first as a "flower" and his role as a "fly on the wall," a purely objective observer, the documentary filmmaker. And it's true that at times he appears to check all of his human superiority and macho male attitude; but there are other moments in this narration—and in scenes used to good effect by Herzog—where Treadwell is anything but humble, where he talks about dominating the bears and showing them that he is the boss, the true apex predator.

> *Occasionally I am challenged. And in that case, the kind warrior must, must, must become a samurai. Must become so, so formidable, so fearless of death, so strong that he will win, he will win. Even the bears will believe that you are more powerful. And in a sense you must be more powerful if you are to survive in this land with the bear.*

In the latter part of his monologue, quoted above, we see him warming up to the part, getting into character, assuming his more subjective role as the alpha male, until he's almost frothing with angst, and suddenly talking about himself in the third person and about how, if pushed or confronted, he would morph from the "kind warrior" into a "samurai" warrior, and become all-powerful, God-like, and master of the beasts. It's a bizarre and fascinating transformation, as if you're watching the monologue of a mad genius.

Herzog, creating his own interpretation of Treadwell's cinematic memoir, front-loads these scenes in *Grizzly Man* so that Treadwell's death is all the more narratively satisfying; but one of the more troubling questions that Treadwell's monologues and the movie raise concerns the role of the ego in wildlife conservation and preservation. Do you have to be a mad genius to get people to pay attention?

Many people considered Treadwell to be a risk-taking charlatan, a scam artist and nut job who didn't really care about anything but

himself. To many he'd crossed a line that shouldn't be crossed. He
was crazy and selfish. Herzog describes "Timmy" as having a "natural
tendency toward chaos." Many others believed he was misunderstood
and saw him as a conduit for wonder and love, the "kind warrior" he
often claimed to be. Treadwell was a contradiction, a charismatic barbarian. One thing for sure, he drew attention to himself and, for better
or worse, to grizzly bears.

Treadwell was even known to act like a bear upon meeting other
humans in the bush. He would "woof," huff and stomp, imitating
a bear's behavior if threatened. Foxes were his constant compan-
ions, appearing regularly in his film footage, acting almost like his
domesticated dogs. In some ways, Treadwell had achieved what David
Villalobos insisted he was after when he leaped into that zoo cage.
He'd crossed over and become one with the animals; he believed the
bears were his friends, that they understood each other. He believed
they had a deep connection. But the truth remains that, no matter
how close he came to being a bear, Treadwell was still a human, still
only a visitor to their world.

Bear biologist Larry Van Daale says of Treadwell's mission:

> . . . *when you spend a lot of time with bears, especially when*
> *you're in the field with them day after day, there's a siren song,*
> *there's a calling that makes you wanna come in and spend more*
> *time in the world. Because it is a simpler world. It is a wonderful*
> *thing, but in fact it's a harsh world. It's a different world that*
> *bears live in than we do. So there is that desire to get into their*
> *world, but the reality is we never can because we're very different*
> *than they are.*[5]

---

5. Taken from transcription of *Grizzly Man*.

Treadwell clearly didn't see himself as so different from the bears, and he certainly felt a spiritual connection to the animals as well; and thus he becomes a touchstone for a consideration of whether human–animal spirituality is fundamentally selfish, something that cannot be fully shared with someone else or with an audience, can perhaps only be approximated through language. Is this spirituality something we attach to them, making living metaphors of animals that are wondrous in their own right, without our attempts to make them into something beyond their nature? Or does it exist in the animal as well, outside of our human subjectivity?

Spirituality is typically as unique to an individual and subjective as pain or happiness. It's difficult to measure objectively, and spirituality is often imposed by humans on a landscape or on animals with nearly colonial zeal. To name something is to own it. And Treadwell named most of his bears. He renamed the landscape where he stayed. He claimed it all as his own spiritual landscape like an explorer planting a flag in a new world.

Treadwell's ultimately fatal dance with the bears makes for a great story—whether you think he is crazy or not, whether you even like him, it is hard to look away. So Treadwell kept going back, for thirteen years, taking his cameras, filming the bears, talking to them in his high-pitched singsong voice as if they were giant emotionally-stunted toddlers in a grassy Alaskan preschool classroom.

Treadwell then took his films back to actual classrooms and talked to children about bears and about conservation. Kids loved Timmy and his stories about the bears. However, one of the great ironies in Treadwell's mission is that the grizzlies he filmed were not, in fact, terribly endangered. The bears lived in a remote part of Katmai National Park, accessible only by airplane or boat, and surrounded by wilderness. Very few humans ever visited the area and certainly not the quantity of "poachers" that Treadwell believed were threatening the bears. The grizzly population there was, by most accounts, healthy and thriving, regardless of Treadwell's influence.

The bears, though, and Treadwell himself became actors on a safe stage, participants in a larger drama he crafted to highlight issues in grizzly bear conservation. They played the part of threatened and endangered innocents, while Treadwell cast himself as their protector, the "kind warrior," watching after them. He was a storyteller who had trouble separating the story from reality—and thus he became the perfect subject for Werner Herzog.

To see Treadwell interacting with grizzly bears *is* sublimely strange and compelling. Herzog couldn't look away, saying in an interview that he used much of Treadwell's own film footage because it was perfect and beautiful. Herzog found that some of Treadwell's "throwaway" scenes and takes were filled with hauntingly beautiful images. Herzog has also admitted to his own obsessive interest in bears. They appear in several of his projects; but he differs from Treadwell in one fundamental way, something he talks openly about as the narrator in the film:

> *What haunts me is that in all the faces of all the bears that*
> *Treadwell ever filmed, I discover no kinship, no understanding,*
> *no mercy. I see only the overwhelming indifference of nature. To*
> *me, there is no such thing as a secret world of the bears.*

Herzog found no spiritual connection to bears, unless there is spirituality in facing indifferent violence. If the bears were palliative for Treadwell, it was only because he saw them that way. They were, for Herzog, a natural placebo for the chaos that plagued Treadwell's soul, a solipsistic vessel for exorcising demons. In other words, while Treadwell may have considered himself "one with the bears," the bears didn't share his enthusiasm and didn't participate in his story. They didn't give a shit about Timothy Treadwell. He was always one hunger pang, one moment of boredom away from being dinner. But Herzog still understood the value of Treadwell's story, saying, "it is not so much a look at wild nature, as it is an insight into ourselves, our nature. And that, for me, beyond his mission, gives meaning to his life and death."

LIKE THE CRAIGHEADS AND other conservationists, Timothy Treadwell used film as a way of educating the public as well as to raise money to fund his adventures. He was, above all, a filmmaker and an actor. He was no bear biologist or trained naturalist. He was an entertainer. Herzog shows us some of the repeated takes Treadwell recorded, and in these, we see a man constructing himself in the moment. We see an actor, a shape-shifter, someone who moves between identities and roles, everything bleeding through the boundaries. We see a conflicted, lonely, and damaged person fighting his own demons. Fighting and ultimately failing.

Eventually Treadwell became so confident in his ability to live in that penumbral zone between human and animal, so comfortable or perhaps intoxicated by his role in the story of the bears, that he also brought his girlfriend Amie into the bears' home, to a place he called the "Grizzly Maze," a much more dangerous place than the "Grizzly Sanctuary" location where Treadwell spent the early part of his time in Alaska.

Amie was not as comfortable as Treadwell around the bears and did not feel the same bond that he felt. We know this from Herzog reading her journals and telling us about it. In all the hours of footage Treadwell shot, Amie only appears twice, faceless, her identity obscured by sun and shadow, or turned away from the camera. When I watched the movie, I couldn't help but think about my own girlfriend in Alaska and her general frustration with my excitement over being in bear country, being in danger, and living a little lower on the food chain. To her it wasn't romantic or spiritual or exciting. It wasn't humbling. It was just fucking scary.

THOUGH THERE'S NO PHOTOGRAPHIC record of the fatal encounter, no eyewitnesses, we know that on October 5, 2003, a large grizzly bear killed and ate most of Timothy Treadwell and Amie. All they found of Timmy was his head, connected to part of his spine, his right arm and hand, with the watch still on his wrist. Everything else had already been eaten or scattered into the dense Alaskan brush. Amie's remains were found nearby, partially covered in a pile of dirt and leaves, suggesting the bear had filled up and was saving her body for later. Most of what we know of Timmy and Amie's deaths is pieced together from Treadwell's journals, his film footage, and an audio recording of the fatal attack.

Would-be rescuers who arrived on the scene shot and killed a large male grizzly, dubbed "Bear 141" according to the dead man's notes, a bear Treadwell hadn't even bothered to name, perhaps because he knew he couldn't control or contain this animal. When they cut open the bear, they found evidence of human remains—fingers, limbs, and flesh. From the campsite they also recovered a video camera that contained a gruesome six-minute audio recording of the attack. Or perhaps the right word to describe it is "grisly." It stays with you . . . or at least I imagine that it would be hard to leave behind, to get the sound of it out of your head.

One of the more powerful scenes in *Grizzly Man* finds Herzog himself sitting in front of Treadwell's longtime friend and former girl-friend, Jewel. The camera peers over Herzog's shoulder, revealing only a thin profile of his face at the edge of the image, and focuses its objective gaze on Jewel's face as she watches Herzog listen to the recording of the attack with a pair of large headphones.

In a documentary film, there's something very odd about this choice. It is clearly not cinema verité, where the camera functions as an impassive, objective fly on the wall, but instead a film where our narrator is also the director and a character, a witness to things that we cannot access through the film. The movie offers up Herzog as a surrogate for our morbid curiosity, a vessel for the violence and horror. We watch Jewel watch him listening in an intoxicating loop of subjective, mediated witness.

Herzog tells her that he can hear Treadwell yelling at Amie to run away, to get away. He's screaming at her to run. Herzog puts his hand up to his face. Jewel gasps. And in the silence, my mind fills the gaps left in the story, just barely approaching something close to the experience. I can't quite get all the way there with my imagination. It's too much, too hard to contain—perhaps too real.

There is no other sound during this scene in the film. All you're given is the silence of Herzog listening to Treadwell's death. When he hands the headphones back to Jewel and tells her she must never listen to the tape and never look at the autopsy photos, he also tells her the tape will be "the white elephant in the room all your life."

It's such an odd thing to say. *The white elephant in the room all your life*—which I suppose would be a gift you don't want that you don't talk about but you still feel its presence in the margins of your everyday life. But there *is* a peculiar wisdom in this statement. I want to laugh and cringe and, against all cultivated values, when I watch this scene, I want Herzog to hand the headphones to me. I want to clamp them over my ears and ask him to rewind, to take me back to the attack—not in the midst of it, but to the just-before time, the in-between wonder and horror.

Can you hear the line crossed, the moment when there is no turning back? What does it sound like? I can almost hear Treadwell's high-pitched voice, frantic, shrieking at Amie, trying everything he can to get her to run and hide, to get away. He knows he's going to die. But my imagination can only take me so far into that experience; and though I know I shouldn't, I want to listen. And it's true that now you can find online what claims to be the actual audio recording of the attack. I could listen with one click. But I don't. It's not the grisly reality of the attack that most interests me, but the ecstatic and imagined reality. I'm interested in the mediated truth of that scene more than I am in the actual recording. I don't want to listen. I want to watch Jewel watch Werner Herzog listening.

HERZOG HAS SAID IN an interview with *Harper's Magazine* (and in other interviews and lectures) that what he's seeking in his documentary films is the "ecstatic truth," one that is shaped not only from facts but also from fabrication and imagination. He's admitted in the past to scripting and staging certain "real" scenes in his documentary films, even to using actors instead of the actual people.

It's pretty clear in *Grizzly Man* that either the coroner is an actor or he is reading lines that he has rehearsed to play the part. Throughout the movie, in every scene, he seems to be overacting, saying everything with more clarity and enthusiasm than seems natural. At the end of one of his scenes, staged in the morgue, with what appears to be a body on a table, hidden beneath a plastic sheet, the coroner delivers his final line and then looks off-camera at Herzog, as if for approval.

Perhaps there is none to be found, but the camera lingers in this moment for a beat or two longer than seems appropriate. The whole scene feels bizarrely amateur. The coroner's hands drop to his side and his posture relaxes. It seems messy and strange, like an editing error, something that should have ended up in a trash bin of cuts. But Herzog leaves it in, leaves us the viewer suspended in the uncomfortable space of not knowing how much of it is real and how much is scripted, rehearsed, perhaps even fabricated. It's an oddly seductive state of not-knowing; and what I've come to believe is actually a kind of finely-crafted and principled confusion.

And that confusion leads us back to another question: Did Herzog actually listen to Treadwell's death or did he just act like he did? Does it matter? The scene doesn't last for the reported six minutes that the recording covers. Perhaps he only listened to part of it. Perhaps he listened to all of it and then edited the scene down for the film. How close did Herzog actually come to those moments, to the reality of that *white elephant in the room*? He did not, as Alejandro Iñárritu did in *The Revenant*, decide to do some kind of recreation of the attack, which would have been a different kind of ecstatic truth, one that's perhaps too easy and simple, too melodramatic and sensational. It would have been too fictional, too "realistic" instead of being *true*.

Something about our mediated witnessing of Treadwell's death, that strange experience of watching Jewel watch Herzog listen (or pretend to listen) to the recording, invites us as an audience inside the experience in ways that are much more nuanced and complicated than a more traditional recreation or first-person, purely factual testimony could achieve. It creates room for us to use our imaginations, to speculate and wonder. It simultaneously creates certainty and uncertainty; and this is, I believe, at least part of what Herzog means when he talks about the ecstatic truth of something. And while I might argue that Iñarritu comes close to a similar kind of truth through his artful recreation of a bear attack, I understand what Herzog means.

The ecstatic truth in these stories is a truth that's bigger, more wild and vibrant, more dangerous and seductive than a truth shaped only by facts or only by one person's interpretation of those facts—which is not to say that it is a lie, but only a truth that's harder to pin down and label; and I've come to believe that stories of animal attacks—even ghost stories and monster stories, and perhaps all stories we tell each other again and again—are shaped by such ecstatic truths, because it is these truths that put us into a state of sublime confusion, a state from which true knowledge of self and the wider world can emerge.

////

IN THE EARLY FALL of 2009, my wife and I took our two kids camping in Sequoia National Park. We'd just come out of one rough patch in our lives and marriage and were trying to do more things together as a whole family, working hard to try and stave off the impending end, our own little *white elephant in the room.*

This trip happened to be during a particularly active bear season in the park, and we saw warnings everywhere about their presence. Despite the risk, we remained undeterred, and I was even a little extra-excited. Early snows had pushed the bears down from the higher elevations in search of food, and they were desperate to pack on the pounds for winter.

Our campsite was a patch of dirt next to a picnic table perched on a hillside and surrounded by other campsites with little space between us. I felt like a termite clinging to a mound. We kept our food in a locking metal bear-proof box and made sure we didn't have anything with an odor in our tent.

After setting up camp, I loaded our daughter into a backpack carrier and the four of us set out on a short hike from the campsite up to a main road and then another area with trails. We weren't more than a few hundred yards away from the campground, winding our way through a clear-cut forest on a dirt path, when we saw our first black bear. He was up ahead of us, maybe a hundred yards away, off the trail, and he looked big, at least three hundred pounds. We all stopped and watched him for a few seconds and it was exhilarating. I wanted to get closer, so I took a few steps toward the bear.

"What are you doing?" my wife asked.

"I just want to see it better."

I admit it. I wanted to get closer to the bear, to something sublime. It was like an ache. I thought again of our trip to Alaska and our near-encounter with that bear; and I thought about how part of me regretted not getting closer. Part of me believed it was like when I visited the zoo with my daughter and we could just walk right up to the edge.

"Um, no. Seriously," my wife said, "come on. Let's go."

Behind me, strapped to my back, my daughter chattered and pointed at the bear. She tugged on my ears. My son looked up at me with concern on his face.

"Really?" I implored. "You guys don't want to see the bear?"

"We can see it from here, Daddy," my son said.

We turned around and hiked back to camp. But later that evening my son and I were sitting at the picnic table as dusk settled over the campground, and we heard something in the distance, out there in the half-light. It sounded like something walking, moving slowly, crunching leaves and twigs. And it was coming closer.

I turned on our flashlight, playing the beam down the hillside from our campsite, moving it back and forth until I caught the glow of two eyes and the black form of a bear. My son and I stared and watched it coming closer, the bear shuffling along toward us, his head wagging back and forth. My daughter was in the tent with her mother. I wasn't sure what we should be doing. I was about to yell or throw something, make a ruckus to scare the bear away. But for a moment, we dwelled in that space of awe just before fear sets in. It's a brief space, a transient state of being, but one that is undeniably seductive.

Then before we really had a chance to be afraid, the space was interrupted. Out of nowhere it seemed, several men appeared, wearing headlamps and carrying guns. They carried big bright lights and one of them seemed to be holding some kind of tracking equipment. It was the Bear Suppression Unit, a special team of park rangers trained and charged with protecting park visitors from hungry black bears.

My son and I watched as they shined their lights on the bear and blasted it with beanbags or rubber bullets, something nonlethal I assumed (since we were surrounded by tents and cars and people). We heard the bear crash off into the brush, fleeing back into the dark. And just as quickly as they appeared, the Bear Suppression Unit disappeared, leaving my son and me to stare at each other in the waning light.

"That was crazy," I said.

"Yeah. Let's go to bed."

Later that night as my family slept, I lay awake, listening to the night sounds, my mind spinning and racing as it often does. It wasn't fear so much as curiosity and overstimulation that kept me awake—as if I'd taken a bump of some drug. And it was in this heightened state of awareness that I heard the bear again. I listened as it approached our campsite, the sound of its soft paws plodding on the dirt, and I felt both afraid and curious.

More than anything, I wanted to unzip the tent and see the bear up close. I wanted to smell its musky odor; but I just lay there listening through the fabric walls. And the bear didn't waste any time. It slammed its paws into the door of the bear-box, making a loud racket, but then quickly moved on, looking for another mark, and I was left in the ringing silence, wondering if the bear would return, hoping both that it would and that it would not.

The next morning, as their mother slept in the tent, I told the kids about the bear visiting our campsite. I told them about him slamming into the metal box, trying to get at our food, and how, unsuccessful, he'd left, roaming on into another campsite.

Their eyes widened with wonder and they asked me to tell them the story again. And again. And somehow between my telling them and their retelling of the story later to friends and family, it changed. In their memory, one crafted from my telling them what I'd only heard in the dark, manufactured from my mediated witness, the story shifted and the bear not only banged on the food box but also broke into it and ate all of our food.

To this day, my children are convinced this happened. They can sometimes even conjure up corroborating images—chip bags torn, crumbs scattered, tattered boxes of food. They argue with me about the story, knowing the ecstatic truth they possess is bigger, wilder, and more fun than the facts; and at some point, I have to agree. To this day we tell the story of the time the bear broke into our food box and ate everything and we all know what it means.

THE CALL OF THE VOID
(OR PSYCHOLOGY
OF THE LEAP)

*And while we watch the animals in their joys of being, in
their grace and ferociousness, a thought becomes more
and more clear. That it is not so much a look at wild
nature, as it is an insight into ourselves, our nature.*

—WERNER HERZOG

1.

THE YEAR WAS 2007. Always, it seemed. And I was waiting for an elevator. Or waiting for something else. I stood on the ninth floor of a hotel in downtown Atlanta, staring down into a vast central atrium strung with orange cloth banners. And I wanted to jump.

I'd come to town for a writer's conference, staying in a neighboring hotel with an old friend. We'd been in a bad way the whole time. There wasn't much to do outside of panel discussions of things we didn't care about or understand and readings we didn't care about or understand. And both of us were struggling to find something like happiness in our lives. I wasn't even a year into my first full-time tenure-track teaching job and already wondering if I could hack it.

My daughter, who hadn't even been conceived yet, would, in a little over a year, come home to a house that had lost most of its value, sticking us with a nearly $1600-per-month mortgage for a house not worth half that. Within a few more years, her parents' relationship, one that had lasted nearly twenty years, would fall apart under the weight of strains both predictable and unpredictable; and I would be sharing custody of her and her brother.

But in this moment in the hotel, staring into the void, I felt perched at the edge of a vast, cavernous future, my toes creeping over the edge. I wanted to leap.

But I didn't want to land.

Everything around the conference hotel seemed like a chain restaurant or a club with bouncers and a cover charge. So when we weren't sitting at our magazine's table at the conference watching legions of disaffected others stream past us, my friend and I spent a lot of time in our room, drinking from a bottle of George Dickel whiskey and chasing it with tall-boys of PBR.

The hotel public areas were clogged with writers, all of them typing on their laptops or burying their noses in the conference program. The whole place had a bad vibe—desperate, lonely, sad, and self-destructive. Or maybe it was just me. The night before I'd set off the smoke detector in our nonsmoking room when I exhaled a huge cloud of marijuana smoke directly into its digital nose. Convinced that the hotel would be evacuated if I didn't do something (I pictured fire trucks lined up out front, and 10,000 pissed-off writers pointing at me, saying, "He's the one"), I got up on a chair, pulled the smoke detector out of the ceiling, and disconnected the wires. The security guard who showed up later looked at what I'd done and just kind of shook his head. He told me I wasn't supposed to do that. I felt like a child and a fool.

Perhaps this is why I found myself, a day later, waiting for another elevator in what felt like an endless stream of elevator rides, suppressing the urge to jump. It's not that I wanted to die, really. It's that I wanted the release, the escape, and the transcendence of falling. I wanted the ecstatic separation, and I saw myself falling softly, like a leaf, sliding down the orange banners and swinging from them, simian-like and graceful. I would become a creature of air, weightless and ghostly.

"You okay, man?" my friend asked.

"Yeah," I said, gazing over the railing. "It just looks like it would be fun to jump."

He stepped up, looked over and nodded. "Let's go," he said and herded me onto an elevator that would carry us carefully all the way down.

2.

WHEN I READ THE initial story about David Villalobos leaping from a monorail at the Bronx Zoo, I'm sure I wasn't alone in assuming he was mentally ill. Because, really, who in his right mind would do that? Who leaps into a cage with a tiger? Who wants to jump into a hotel atrium?

A second-day headline in the *New York Daily News* read "Zooicide," despite David's repeated insistence that he didn't want to die, that his leap was absolutely not a suicide attempt. But many of us don't know how to categorize a story like David's. It doesn't fit our typical narratives. It's too easy to just say he was "mentally ill" and call his leap a "suicide attempt." It's too vague and meaningless, really, and doesn't address the spectrum of possibilities.

David told the responding officer, Detective McCrossen, that he was testing his natural fear and that it was a "spiritual thing." And there's a truth in the divide between those statements, a kind of second-day story of David's encounter with Bashuta the tiger, that pulls at all my strings and rattles my cage.

3.

WE HAVE LONG SEEN apex predators as tests, challenges not just to our physical bodies but to our spiritual bodies as well. When Daniel was thrown into the lion's den it was as punishment for his prayers and his belief, his loyalty to God over his king, and, as the story tells us, it was his faith that protected him from certain death. When he emerged unscathed, Daniel credited God with intervening and saving him from the lions. But we don't really know what happened in that den, don't know what Daniel saw or felt as the rock was slid into place over the opening. We don't have his memories of those hours or any story to explain how God was able to "shut the mouths" of the lions. This to me is one of the great mysteries of the Bible; but I'm equally interested in knowing not how intimacy with apex predators brought Daniel closer to God, but what that experience revealed about himself and if, in those first frightening moments of encounter, he questioned God's existence. Did he feel shattered? Did he fight or surrender to death and faith?

SCHOPENHAUER, THAT JOLLY PHILOSOPHER of the sublime, might encourage all of us to leap into cages with predators, or at least encourage us to consider the possibility. He believed that if you were to attempt to contemplate certain "contemplation-resistant 'objects' or phenomena that bear a hostile relationship to the human will insofar as they are so vast or powerful that they threaten to overwhelm the human individual or reduce his existence on this planet to a mere speck," then you would be approaching an experience of the sublime; and I think I'd put a tiger into the category of phenomena that "bear a hostile relationship to the human will."

These phenomena are physical and natural (NOT supernatural)—often things such as violent storms at sea, earthquakes, a frozen landscape, or the starry sky at night, perhaps even a den full of hungry lions, or just a cage with a single tiger, a pool and a polar bear, a grizzly bear at midnight—and to come face-to-face with such phenomena is to experience a force beyond rationality, beyond understanding, and beyond God. And Schopenhauer's aesthetic theory—as delightfully spare and harsh and unforgiving as the frozen landscapes he valorized—also argues for the appreciation of tragedy, pain, and suffering (even in others) as perhaps the highest experience of the sublime; which would then seem to carve out a different identity for your local zoo. No longer a sad collection of cages, the zoo now becomes a place where the point is to appreciate the suffering of others, where you can come face-to-face with the sublime.

I believe that this experience of the sublime is what we are seeking when we get up close and intimate with apex predators, or when we can't stop reading stories about people who jump or fall or are thrown into a cage with them. Perhaps the sublime is what Daniel experienced in the lion's den; but if so then maybe he experienced not so much God but the secular dynamical sublime, at least in that ecstatic moment before God's intervention, when it was possible that he would be killed and consumed by lions.

Maybe this is also what Ahab experienced as the white whale devoured him and the *Pequod* sank to the bottom of the sea. And maybe this is what David Villalobos felt in those few fibrillating moments with Bashuta. This is also close to what I experienced as I gazed down into that hotel atrium in Atlanta.

4.

IF WE CONSIDER DAVID Villalobos and Timothy Treadwell or even Stephen Haas through the lens of psychology, we can perhaps begin to see some things a little clearer or come close to something like an answer.

Werner Herzog believed that to look into the eyes of a bear or some other great predator was to gaze at moral indifference, to face the reality that the bear cares nothing for you and is always one choice away from making you into a meal. It is to face your own meaningless annihilation.

Though there seems to be very little research focusing on intentional encounters with predators (imagine the clinical trials!), David Villalobos and Timothy Treadwell, maybe even Stephen Haas, could probably be categorized as "sensation seekers," or "thrill seekers," and (much to my advantage) it turns out there is a great deal of clinical research into the psychology of such individuals. Doctors Seymour Epstein, Frank Farley, and Marvin Zuckerman have each written extensively on the question of why someone would actively seek out a thrill that may possess the potential to kill the seeker.

Research conducted by Zuckerman on "sensation-seeking" personalities suggests that there is a spectrum of sensation seeking from low to high, with the high sensation seekers being the kind of folks who leap out of airplanes or jump motorcycles over canyons, people who are constantly looking for novel, interesting, and high intensity experiences. You know these people. They are the ones who participate in the X Games. They climb aboard massive snow machines and flip them into the air. And perhaps they are also the kind of people who try to get up close and personal with a grizzly bear or a tiger. Perhaps, even, they are the type of people who consider leaping into a hotel atrium. Such individuals thrive on uncertainty and intensity.

Thrill seekers are probably a minority in our population, but they are an attractive and narratively compelling minority. Think for a minute about how many news stories that grab our attention involve one person pushing the boundaries of what's safe or acceptable, one person craving a novel experience, something new and different and more meaningful.

Some have even suggested that our country was founded by "sensation-seeking" adventurers, striking out into the wild unknown with no guarantee of anything but a novel experience and the thrill of discovery. And it is true. We are a nation of thrill seekers who value the leap more than the landing.

5.

THOUGH I'VE SPENT PLENTY of time on motorcycles, skis, and snowboards, and I grew up shooting guns and riding bikes, I never really associated much with the thrill-seeking, extreme-sport crowd. I had too much fear, too many worries over consequences. Mostly I was trying to keep up with my little brother, Matt, who had always been the more fearless, more extreme, more thrill seeking between the two of us. I chose the safer, more acceptable team sports activities of football and basketball. Matt chose BMX, where he excelled rapidly and eventually became one of the top riders in the Missouri–Kansas region, collecting a shelf full of trophies, one of them taller than me.

I saw myself as risk-averse, overly worried, and much more interested in being popular, which is probably why I decided to be a jock and play starting roles on both football and basketball teams. The differences between my brother and me were hard to negotiate sometimes, particularly since they extended beyond just sports and extracurricular activities to our personalities and to our values.

Matt always claimed to be an atheist, even if I argued that doing so required just as much blind faith as it does to believe in God. One of his life goals—he liked to tease me with—was to own a Jaguar. Not the cat, but the car. He was unabashedly driven to be successful and wealthy and the best at whatever he chose to do. I was driven to retire at an early age. For a while I planned to be a river raft guide. Or maybe a professional fisherman. Matt wanted to build robots that build cars.

I dreamed of hosting a fishing show with my friends, or living in a cabin in the woods. He and I didn't get along for much of our adolescence and didn't really have time to get to know each other as adults.

My little brother was killed in a car accident when he was eighteen, the summer after his freshman year of college. After leaving the movie *Lethal Weapon 3*, Matt's car slid out on a turn and slammed into a tree. He was always going hard and fast, right up to his end; and I've written many times about his death and speculated in the past that what separated us—our ability to deal with fear—is also what allowed him to live louder and more ecstatically.

It's too easy, though, to say it was some thrill-seeking drive, some part of his personality, that ultimately killed my brother. It's never that simple. I probably would have put myself on the lower end of the sensation-seeking spectrum and then placed my brother at the high end. But I'm beginning to understand now that it's not such an easy binary between us, and that ultimately we were closer than either of us realized—each of us chasing the sublime and the ecstatic in our own way.

In his research, Farley tries to refine the distinction between the different kinds of thrill seekers and identifies what he calls positive and negative "Big-T" personalities, where the positive Big-T's might be characterized by disinhibition and impulsive behavior and involved in thrill-seeking, risky pursuits. The interesting thing is that Farley suggests these people also seem just as likely to be involved in creative science or artistic endeavors.

Farley identifies Albert Einstein as a Big-T positive; whereas negative Big-T's tend toward aggression, violence, or other impulsive antisocial behavior with little regard for the consequences. These are your alpha male types; the guys who seek out fights or other intense physical encounters, the men who want to prove their dominance over others. These are the guys I don't want to become.

Mike Tyson might be a Big-T negative, or at least that's how most people see him. And pop culture has given us a history of characters who embody the binary and the contradiction between Big-T positive and Big-T negative—from Dr. Jekyll and Mr. Hyde, to Bruce Banner and the Incredible Hulk, or Hank McCoy and Beast—where the positive side is often a man of science and the other is an animal, a monstrous creature who only comes out during threat situations. The point of these characters is to exaggerate the "animal" side of man, to suggest that in each of us is a beast waiting to manifest itself, that there is an uncontrollable, uncivilized Big T "wildness" in every man, perhaps in every person, that each of us contains a savage within.

6.

In ADDITION TO BEHAVIORAL differences, it also appears that there may be both physiological and biochemical differences between low sensation seekers and people like David Villalobos or Timothy Treadwell. Put simply, their brains just work differently. When exposed to intense stimuli (in one case a series of graphic images in a study that sounds eerily like *A Clockwork Orange*), the Big-T's brains were active in areas associated with addiction response and dopamine regulation. In other words, when faced with risk or threat or uncertainty, there really is a biochemical rush, a high that is more intense and more seductive, and perhaps even addictive for Big-T types. They get off on it.

If you watch some of the footage of Timothy Treadwell, he looks a little crazy. Or like he's high on some exceptionally good drugs. He gets REALLY excited—hyper-verbose and euphoric—and then quickly he becomes REALLY depressed, despondent, and cynical, even hopeless and openly angry. He rapidly vacillates between over-the-top macho aggression and almost childlike wonder and innocence. He slips in and out of performance and reality, between a role he's playing and himself. At times he looks a lot like a vain, coked-up charlatan, while at others he seems like a peaceful, generous, nature-loving hippie. Probably he was a little bit of both; and, for whatever reason, the grizzly bears seemed to magnify these poles in his personality.

Treadwell reminds me a lot of Mike Tyson. They embody contradictions. They both talk in an oddly compelling high-pitched voice.

They both seem incredibly self-conscious and, at times, painfully inse-
cure, but also, paradoxically, somehow dynamic, self-assured, and
attractive. They are charismatic barbarians—slightly wild but undeniably magnetic mixtures of man and animal. They are ecstatic manifes-
tations of strength and vulnerability. And they both choose to jump
into the ring with apex predators. Again and again and again, even if,
most of the time, each is his own worst predator.

7.

SOME RESEARCH INTO THRILL-SEEKING personalities suggests that getting intimate with apex predators is, perhaps not surprisingly, correlated to higher levels of testosterone. Thus it makes some biological sense that my girlfriend wouldn't have shared my desire to get up close to a grizzly bear in Alaska. It makes biological sense that Treadwell's girlfriend Amie also wouldn't feel the same rush that he felt. Perhaps her brain didn't work the same way. Her body didn't flush with the same biochemical flood of sensations. Mostly she was just scared. Mostly she just wanted to be somewhere else.

The thing that I found most compelling was the association Farley makes between thrill seeking and art. I don't know that I ever considered writing to be a kind of "sensation seeking," since it is obviously very different from skydiving or hang-gliding, and pretty far away from jumping a motorcycle in the air; but there is a kind of rush to the experience, a momentary thrill that comes from writing that pulls you back again and again, kind of like a drug.

There is also Schopenhauer's aesthetic appreciation of tragedy, which somehow seems to legitimize so much of how I spend my free time, obsessing over the pain and suffering of others. And I believe good writing is driven in large part by the constant craving for new experiences and fueled by a drive toward the novel and sublime.

Curiosity is the engine of most great art, and the best writing spawns from a restless soul. But whatever you feel as a writer during the writing process is mostly artificial, a recreation or second-day story of the actual experience. This doesn't mean that such recreations are any less meaningful, or the sensations they induce any less powerful. In fact, they may even be *more* meaningful, especially if we can accept that "meaning" is mostly a human invention, an artifice that we agree to accept as a substitute for the real experience. But I somehow found it edifying to read about writing or art as a kind of thrill-seeking endeavor. It brought me closer to my brother and to these other men.

My efforts to relive the experience of Stephen Haas—my defensive imaginings and preoccupations with people like Mike Tyson or Timothy Treadwell—could be, according to some of this research, its own kind of experience of the sublime. The page, thus imagined, becomes also a kind of void, and writing a spiritual leap into the empty white space, a leap that frightens and paralyzes a great many people, and one that still fills me with equal parts terror and wonder.

8.

MY EX-WIFE USED TO refer to me as an "edge stander," meaning that I liked to get right out on the edge of things. I'm not afraid of heights but rather compelled by them. I feel it on mountain cliffs, bridges, and tall buildings. I feel it sometimes with an essay—that teetering on the brink of something.

Before he died, my brother had taken up skydiving. He did it for fun, just something for kicks on the weekends. He loved to leap, and I'm quite sure he never felt fear in the same way that most of us feel it. He could just switch it off and jump.

I wonder how it feels to be perched at the door of an airplane waiting to jump, or on the edge of a zoo cage staring down at an apex predator, and if it's anything like what I feel when I gaze down at an empty page.

Perhaps there is something about a cage, that like a cliff inspires one to suspend the normal rational consideration of consequences. Perhaps a cage can call to you as well, and death at the claws and teeth of a predator becomes the same promise as a landing, that inevitable consequence you don't want to acknowledge because it ruins the thrill. A tiger is also a kind of void, after all, a kind of existential nothingness, a vessel of your own annihilation.

Recently I took my kids hiking in the Sierra Nevada mountains, to a place called the Fresno Dome, a massive granite dome sheared off on one side and curved like a half-sphere on the other; and as we approached the summit, I felt something I'd never felt before, some-thing entirely different from what I felt a few years before in that hotel lobby in Atlanta when my daughter was just an idea.

She'd now grown into a fearless and tough child, wise beyond her years and full of personality, not to mention an "edge stander" in her own right. As I trudged up the rocky path, she bounded up ahead of me, moving quickly, and disappeared over my horizon. I called out to her. My guts quivered and my legs felt weak. My heart fluttered like a bird trying to take flight in the cavity of my chest. And as I crested the rise and found her, not far away, smiling, just a few feet from the cliff, I felt something strange and discomforting.

I realized later that it was vertigo I'd felt. I'd suddenly, quite unexpectedly, encountered a fear of falling, and I didn't recognize it. My daughter had changed everything. The void was not calling to me. It was screaming at me to stay back. Suddenly I'd been terrified of the height, afraid to get too close to the edge. The panic rose like a tide and I couldn't hold it back. I grabbed my daughter's arm, gripping her tight, and pulled her back. *Not yet. It's too early for you.*

9.

PERHAPS YOU, TOO, HAVE felt the sudden overpowering urge to leap from high places, to jump from a bridge or cliff or tall building; maybe that momentary desire to drive your car over the edge, the seductive pull of vertigo as your car hugs the edge or as you're standing for a photo opportunity with your family at some scenic overlook; maybe you know intimately the urge to leap from towers or monuments—but not necessarily the urge to land. I'm not talking about suicide ideation, but something else entirely, something the French call *l'appel du vide*, or *the call of the void.*

Or perhaps you simply can't stop reading and writing about people who make such leaps.

It is not a death wish but instead a kind of urge toward ecstasy, toward the extreme release and departure, an urge to escape into nothingness; and perhaps it is also a thrill-seeking urge, a Big-T kind of drive and desire. It's not even the urge to fly or to be released from the bonds of gravity, but instead to be suspended within them, to live, if only for a moment, in between action and consequence.

Once I did leap like my brother—into the void. One hundred and fifty feet up, standing on the edge of a crane basket, looking down on a parking lot filled with Grateful Dead fans in Mountain View, California. There was no shark pit, no cage below, and no tiger waiting; but there was a man dressed in a red dancing bear costume and the bored attendants just doing their job. The psychedelic mushrooms

I'd eaten hadn't really started to kick in yet as one attendant secured the bungee cord to my harness and instructed me to stand on this tiny piece of metal. He shut the gate behind me and started counting down from ten . . . nine . . . eight . . . seven. I can still see me falling, suspended there between what was and what would be, between action and consequence, existing in the liminal space between the leap and the landing. I think part of me has been trying to get back to that space ever since.

PART THREE

CHARISMATIC BARBARIANS

MANIMALS
IN CAPTIVITY

*There are very few people who are going to look
into the mirror and say, "That person I see is a
savage monster"; instead, they make up some
construction that justifies what they do.*

—NOAM CHOMSKY

1.

DAVID VILLALOBOS LEAPT INTO that tiger cage at the Bronx Zoo and tried to cross over. He wanted to become one with a tiger, to bridge the divide between man and animal; but for me his leap made him more than simply a sensational subject and instead a lens and a looking glass.

My father often read to my brother and me from Rudyard Kipling's *Just So Stories* or *The Jungle Book*, where we fell in love with the mongoose Rikki-Tikki-Tavi and, most of all, Mowgli the wolf-boy and his friends, Bagheera, a black panther, and Baloo, the sloth bear. Mowgli was perhaps the first crossover character or charismatic barbarian who captivated my attention—a boy, raised by wolves, living amongst the beasts, away from civilization, running from the fierce tiger, Shere Khan, the "man-killer." The movie adaptation of *The Jungle Book* was also a big favorite and one that I watched over and over again as a child, and then again as an adult as I passed on the story of Mowgli and his mates to my own children. These stories persist across time, crossing generational gaps, becoming mythic and archetypal.

My childhood interest in characters like Mowgli was also nursed along by a string of television shows and movies that aired in the late '70s and early '80s. These books, movies, and TV shows featured characters who moved between the two worlds or who lived on the edge between human and animal, wild and civilized; and these characters were some of my earliest and most appealing pop culture influences.

I'm still not sure what it says about me that, as a young boy, my heroes were half-wild men, hermits, castaways, and barbarians; or that my plans for the future consisted mostly of living alone in the woods, and my models for friendship were socially awkward bears. But I'm pretty sure I wasn't alone in idealizing these relationships, nor am I alone now in trying to puzzle out their meaning some thirty years later, as I settle into middle-age fatherhood and watch my own kids find heroes and villains in pop culture.

Sometimes I wonder if David Villalobos had heroes like me. Did he, too, imagine a reality where he could exist between the worlds of human and animal, perhaps where he could live with apex predators in harmony, accepted as one of them?

Mowgli and Sam Gribley and other charismatic barbarians promised the sort of reality that must have drawn people like Timothy Treadwell or Christopher McCandless into believing such a life was possible. They promised a gentler path, a journey of harmony between man and animal that didn't inevitably end in violence and savagery.

2.

THE 1974 TV MOVIE and subsequent 1977–78 TV series *Grizzly Adams*, starring Dan Haggerty, featured as its protagonist a "mountain man" with a truly epic beard and mythic presence; but more significantly for me, his best friend and partner was an enormous grizzly bear named Ben.

Charles E. Sellier, the man largely responsible for cultivating the myth of Grizzly Adams, admitted in a *TV Guide* interview that the original movie he produced was created based on market studies and audience tests that showed people loved "stories about men and animals in the wilderness; that bears were favorite wilderness animals; and that grizzlies were the favorite type of bear."

A creation of the culture, a product marketed to a buying public, Sellier's character was also based on a real-life man, John "Grizzly" Adams—a hunter, trapper, mountaineer, and collector of animals who was once called the "Barnum of the Pacific." The movie and TV series also worked to directly contradict a changing perception of bears as monsters by featuring a gentle giant, Ben the bear, as a kind of co-protagonist. Contrary to the terrifying image of bears in *Night of the Grizzlies* and elsewhere, Ben was kind and loving, loyal and protective of his friend. Ben was a throwback, a benevolent force, but he was also a force to be feared.

The movie and TV show were based on the novel *The Life and Times of Grizzly Adams*, written by Sellier, which itself was heavily influenced by a biography of Adams by Theodore Hittel published in 1860, *The Adventures of James Capen Adams, Mountaineer and Grizzly Bear Hunter of California*, a book that did much to establish the mythic presence of Adams in the history of California and the west. Hittel wrote the book after he'd met Adams at his San Francisco museum and, captivated by his character, spent a great deal of time listening to and recording the trapper's tales of adventure. There have been several other books since, most of them agreeing on the rough outline of Grizzly Adams's life.

Though he adopted several different monikers throughout his travels in California and undoubtedly fabricated, or at least embellished, parts of his personal story, most sources seem to agree that Adams was a New England cobbler by trade and a trapper, who was once nearly killed by a tiger owned by his employer. This odd fact seems especially significant, given that Adams would eventually die from a festering head wound inflicted during a wrestling match by one of his captive bears.

Moving to the High Sierras of California with the forty-niners, Adams had reinvented himself as a hunter, trapper, and leather-smith, as well as a fierce and formidable bear-fighter. He made a living by providing shoes and clothing to miner camps, and by trapping and hunting. Adams captured more bears and other predators alive than most trappers did by shooting them or poisoning them, and he attained mythic status as someone who could tame and control even the most savage beasts.

One 1976 article in the *Yosemite National Park* magazine, written by Dean Shenk, gives a hint at the mythic presence of Adams: "After spending close to a month within what is now Yosemite National Park, Adams headed home with numerous hides, 'Ben Franklin' (Adams's longtime friend, a bear who sometimes let Adams ride him with a saddle), two wolf cubs, five cougar kittens, and a pair of fawns."

In 1856 Adams came down from the mountains for good and opened his Mountaineer Museum on Clay Street in downtown San Francisco, which he would later move and rename the Pacific Museum. The museum featured both live and taxidermied animals that Adams had caught or killed in the Sierra Nevada Mountains. Adams's animal menagerie was a veritable live-action wondercabinet that displayed, by some reports, nine bears, including the 1500-pound behemoth, Samson. Adams's museum also featured elk, mountain lions, eagles, and other smaller birds and mammals.

Some writers have even speculated that Adams's San Francisco museums, and the traveling menagerie he eventually took to the East Coast as part of P.T. Barnum's circus, became a major inspiration for the rise of public and private zoos in America, including the Bronx Zoo in New York, which opened in 1899, and the Central Park Zoo, which was originally an informal menagerie of donated and abandoned animals.

Grizzly Adams was California's real-life man-of-myth, our own Paul Bunyan, and a figure who looms large over the history of this place I've come to call home; but he was also a '70s TV star, a pop culture hero of my childhood and a role model. He was both real and unreal, natural and manufactured, like a buckskin superhero, a pioneer Beastmaster. I'd been raised in the culture of Boy Scouts, taught to always carry a knife and be prepared. My father had schooled me on how to build a fire and gut a fish or clean a bird, how to find my compass directions, and when to follow a stream. Grizzly Adams became a kind of embodiment of all that I wanted—a bearded survivalist with a bear for a buddy and a TV show. He was famous for his rejection of fame, his escape from the trappings of civilization.

Some days, at my writing desk, I can still see the mythical Grizzly Adams now off in the distance, crossing over the horizon into reality—a thick-bearded man, wearing the hides of bears, riding an enormous grizzly named Ben Franklin, a rifle slung over his shoulder, a Colt pistol on one hip, a bone-handled hatchet on the other. He has become one with the animals, an accepted member of their tribe. They trust and respect him. Unlike the real Grizzly Adams he is not dying from a festering head wound inflicted by a bear during a "wrestling match." No, this Grizzly Adams is fully and ecstatically alive in that in-between realm. An eagle perches on his shoulder, songbirds nest in his beard, and squirrels race each other around the trunks of the bear's legs. A family of field mice store nuts in his pockets, their tiny heads peeking up over the rim. He raises an arm to wave, and ravens burst forth from his armpit, circling the sky, cawing to him, while down below, the fawns panic a bit, their hooves beating a staccato dance on the rocks. With a quiet word and wave of his hand, Grizzly Adams tames the beasts and they all follow him like acolytes, down into the Valley of Civilization.

3.

A FEW YEARS AGO, over the Grapevine mountain range, down in a different valley, at Disneyland, I caught sight of another charismatic barbarian, the Beastmaster, waiting in line for the Mad Hatter's teacup ride. And by that, of course, I mean I saw Marc Singer, the actor who played the sword-swinging, animal-loving barbarian, Dar, in the 1982 fantasy film *Beastmaster*. I saw the actor, not the character—though it was, admittedly, a little tough for me to keep them apart in my mind.

I'd come to the Land of Disney with my wife and our son. We'd driven four-plus hours to visit some friends in Hollywood and to escape the murderous San Joaquin Valley heat; and there we all were, nestled into the ample bosom of the happiest place on Earth, queuing up for the Mad Hatter's ride, when I spotted him.

Across the winding maze of happy people, I caught a quick glimpse of the Beastmaster. Brief at first, my gaze shifted away, but I kept glancing back, subtly collecting the confirming details—angular face, aquiline nose, overly large nostrils, broad shoulders, and blondish hair. It took a minute or two for me to register Singer's face and place him in the films I knew; but when I did, I felt that strange satisfying heat of nostalgia settle into my gut.

Singer had skin that looked sort of stretched and tanned like leather; but he was still in good shape, still strikingly handsome. He's only a couple of years younger than my parents and the age was evident in his face, but his body was sculpted and fit, his biceps and pecs

filling out a tight T-shirt. He looked good, better than I imagine I'll
look at his age. To the other people in line, he could have been any
average L.A. guy, just a fit suburban dad or fitness buff with a fake tan
and dyed hair. But I knew the truth. Underneath those normal clothes,
he was the Beastmaster.

I felt strangely happy to see that, if called upon, Singer could still
play the role of a shirtless barbarian. Somehow this seemed to preserve
a small part of my often miserable but occasionally transcendent adolescence. I'm pretty sure Singer was riding the teacups with a much
younger woman, but honestly I was so fixated on him I don't really
remember his companion. He might have been there with his kids.
Or his dentist. Or his mother. Or some kind of Hollywood player. I
didn't really care.

Nobody else in line seemed to recognize Singer. Or at least
nobody asked for his autograph. Nobody said, "Hey, Beastmaster. You
rock!" though I did point him out to my son, who had no idea who
I was talking about and mostly ignored me—whereupon I resolved
then and there to, at some distant point in his future, school the boy
on the finer examples of '70s and '80s barbarian narratives. It seemed
like the least I could do—though now, of course, he's reached his teen-
age years and would undoubtedly find my love of crappy '80s fantasy
flicks to be both sad and embarrassing.

Let's be honest: *Beastmaster* is not an objectively great (i.e., criti-
cally acclaimed) movie, but it was an important movie to many kids
growing up in the early '80s. Though a plot summary doesn't quite
capture the epic quality of the film, IMDB describes it thusly:

> *Dar, the son of a king, is hunted by a priest after his birth, so he*
> *is sent to grow up in another family. When he becomes a grown*

man, his new father is murdered by savages. He discovers that he
has the ability to communicate with the animals, and after that,
Dar begins his quest for revenge in this Conan-like movie.

That line about the "ability to communicate with animals" gets me every time. I grew up watching old Tarzan episodes on Saturday mornings, obsessed with the apeman's ability to live between two worlds so comfortably, captivated by the "idiot savant" traits of the charismatic barbarian.

Raised by wolves or apes or dolphins, Tarzan and the other crossover characters were mutants and freaks, and yet still somehow uniquely adapted for survival. They looked like "humans" and exhibited "human" characteristics, but also possessed both the best and worst of "animal" characteristics—strength, power, savagery, and moral indifference. Most importantly, they possessed the ability to communicate between species. And if I'm being totally honest, amidst the angst and turmoil of the '80s, a painful divorce and the usual adolescent insecurities, I often wished I'd been raised by wolves or bears or some other family of beasts. Some days it felt like nobody understood me and I'd be better off talking with animals.

Beastmaster was marketed with a poster of a tanned and muscled Marc Singer coupled with a tagline, describing our hero thusly: "Born with the courage of an eagle, the strength of a black tiger, and the power of a god," which I've decided is exactly what I'd like etched into my tombstone when I die.

Using his far more advanced skills in Beastmastery than I could ever hope to possess, Dar befriends an eagle named Karak and travels with the aforementioned "black tiger" as well as a buxom "slave princess" and two thieving ferrets, all of them happily internalizing Dar's quest to find and kill the evil "priest," the sorcerer, Maax (with two a's), responsible for the deaths of his parents. It is perhaps worth noting that several of my childhood heroes of literature and pop culture possessed the unique ability to domesticate predatory raptors.

The revenge-fueled barbarian quest was a popular trope in '80s movies and television. Between the Conan movies, *Beastmaster*, *Mad Max*, and the animated TV series *Thundarr the Barbarian*, I consumed many stories of semi-savage, muscled men who traveled with a team of justice-loving outcasts and/or animals through a dangerous, fantastical, and/or post-apocalyptic world filled with mutants and freaks. It was the '80s, after all, and such stories seemed not just inevitable but entirely necessary.

Often these "wild" men, these barbarians, were pursued unjustly or hunted by a persistent threat—a priest or a cop, a general or colonel, a mad scientist or a mercenary. Often there was vengeance. And scantily-clad women. And fighting for justice. And swords and loincloths and stuff. And even Tina Turner and Grace Jones and Wilt Chamberlain and Andre the Giant. It was amazing and ridiculous, pretty much like everything else in late '70s, early '80s pop culture—hedonistic and base and undeniably seductive. The movie makers knew their audience—legions of disaffected young, middle-class white boys who grew up at the tail end of the Cold War (when things got really weird), boys who loved comic books and metal and *Star Wars* and, especially, Princess Leia's gold bikini, boys who harbored dreams of vengeance and epic escape through a fantastical barbarian bildungsroman—boys just like me.

4.

MANY OF MY FAVORITE TV shows and movies from late '70s and early '80s were objectively bad and "short-lived"; perhaps I liked them because I appreciated the transient significance, the lingering resonance of camp and melodrama. Perhaps my appreciation can be explained by a feeling during those years that everything meaningful had a short shelf life, destined as we all were to die in storms of nuclear wind and fire. Perhaps it was my long-running fascination with the one-hit wonder, the flash in the pan, or the earnestly crappy expression of pop culture. I also loved the songs "Mickey," and "867–5309"; so perhaps I simply had bad taste.

The 1977–78 series *Man from Atlantis* was one of these short-lived TV shows starring Patrick Duffy and was clearly based on the comic book character Aquaman. It featured a protagonist named Mark Harris, who had washed up on a beach, suffering from amnesia and sporting webbed hands, sculpted pecs, and some really awesome feathered hair. Harris, believed to be the last known survivor of the lost city of Atlantis, swam like a dolphin with his arms at his side, kicking his legs in unison instead of in a scissor-motion, making his body wiggle and wave like a worm in the water. This, perhaps not surprisingly, was something I tried to imitate EVERY single time I was in a swimming pool as a kid, and still do today just to show off for my kids and pretend that I, too, could be the Man from Atlantis, if only for a moment, in the suspended animation of a hotel swimming pool.

Aside from his uniquely entertaining swim style, Mark Harris also possessed superhuman strength and the ability to communicate with sea creatures—both of which come in handy when you're a mutant freak being hunted by a madman. The story hit a lot of marks for me. But in terms of narrative tension, the show perhaps depended too much on the audience internalizing a fear of this aquatically-inclined madman who, when he wasn't relentlessly pursuing Harris, was creating an undersea kingdom from which he intended to nuke the surface world into oblivion. Again, I didn't see a problem with this. It seemed to be a pretty obvious analogy for the current geopolitical climate.

The show had it all. What's not to like? Superhuman mutants with feathered hair. The clash of science and nature. The blurry lines between man and animal. Nuclear bombs. This was pop culture at its best. Unfortunately the critics, audience, and studio executives didn't feel the same way, and *Man from Atlantis* was canceled after only thirteen episodes; and many members of the target audience—the savage youth of '80s America—believed that our own seasons, our own fragile pilot lives, would be canceled soon enough, wiped out by the cold logic of nuclear war.

5.

WHAT DOES IT MEAN to become one with an animal, to bond with an apex predator in an intimately spiritual and perhaps even physical way?

Admittedly it is difficult to articulate this crossing over. I know it's more complicated, more esoteric and sublime than being consumed by a predator and probably difficult for David Villalobos or anyone else to explain. It is to be consumed but not eaten, to be absorbed but not destroyed. It's perhaps the kind of spiritual connection, or communication between species at which Werner Herzog would scoff and laugh, dismissing the idea with a wave, believing in the deep indifference of nature. But part of me likes to think that this crossing over from man to animal might also look a lot like it did in the 1983 TV series *Manimal*.

Stay with me here.

Manimal featured the protagonist Dr. Jonathan Chase, played by Simon MacCorkindale, a shape-shifting private detective described in the show's voiceover intro as:

> . . . *wealthy, young, handsome. A man with the brightest of futures.*
> *A man with the darkest of pasts. From Africa's deepest recesses, to*
> *the rarefied peaks of Tibet, heir to his father's legacy and the world's*
> *darkest mysteries. Jonathan Chase, master of the secrets that divide*
> *man from animal, animal from man. . . Manimal!*

Manimal exists, for me, as a cultural touchstone, a nexus of meaning and madness. The show's run coincided with the post-apocalyptic TV movie *The Day After*, the Beirut bombing of the Marine barracks, the downing of Korean Airlines Flight 007, the invasion of Grenada, and, ultimately, of my parents' divorce—all of these bombs dropping during the fall of 1983.

Boom, boom, boom, boom.

To say that those long strange months were significant in my life would be an understatement. It was, for me, the season of apocalyptic synchronicity—when all my private hopes and fears seemed to play out on the public stage and onscreen. Terrified of nuclear war and of the Soviets and terrorists, obsessed with mutation and solitary adventures in the wilderness with animal friends, I felt like they were making movies and TV shows just for me. Sometimes the whole world seemed sick, like a bad fantasy movie, and *Manimal* felt like the only cure. I don't think I was alone in believing that, when the inevitable end came, it wouldn't be the meek but the mutants and the manimals who'd inherit the Earth.

I suppose I felt the sort of solipsism that most people feel in their adolescence, but it was also a time of significant self-reflection for me, and for an odd kind of empowerment and acceptance through the glory of bad movies and crappy television, through the revenge travel narratives of charismatic barbarians—those brief, bright heroes of the screen who somehow touched the archetype and spoke to an internal struggle so many of us felt deep in our marrow. The world—all of it—often seemed so serious, so dire and depressing, that we all needed a release, a promise of better times. We needed to feel stupid and barbaric, if only for a season, as a salve against the smarts that seemed required to survive the '80s.

6.

THOUGH HE COULD CHANGE into any animal, Dr. Chase, or Manimal, most often morphed into a hawk or a black jaguar, presumably because these two animals were the best at fighting crime. The hawk could . . . you know . . . fly and see really well and grab things with its talons; and the jaguar was a big predatory cat . . . that mostly just scared the living shit out of bad guys. When the cat showed up, the party was over. Surrender was imminent and predictable. There were few, if any, scenes where the animals actually attacked a human. No vicious jaguar or bear maulings. Mostly Chase relied on the strategic intimidation the animals could offer.

And here's the important distinction in the *Manimal* story: Chase's animal nature, you would learn, was not chaotic, not wild and rampant, but instead a product of rational thought and planning, a willful morphing from man to beast. It was a way to solve problems, to avoid violence through intimidation and performance. Peace through strength—just like Reagan's nuclear and foreign policy. Manimal was more American, more Cold War, than we could even imagine.

At every step, Chase had to think it through and plan his transformation into an animal. It was always a calculated move. Jaguar as gun. Hawk as badge. He chose to be a monster and a menace when it benefited the common good. Chase had succeeded where Bruce Banner and the government couldn't with the Incredible Hulk; he had controlled and weaponized his animal side.

Let's assume, then, that the Hulk is a metaphor for the false promises of the atomic age, a big green manifestation of chaos, a rampaging unpredictable nuclear meltdown. Jonathan Chase, then, exists as a kind of mutative dream, the perfect picture of successful Cold War survival—as if James Bond and the Hulk had a handsome and well-adjusted baby boy named Jonathan. As a general public, we were just lucky that Chase chose to use his mutative powers for good. And we were lucky that cameras were rolling to capture it all.

One of my favorite parts of *The Incredible Hulk* TV show was the mutation sequence, where you watch the slow, seam-splitting transformation of Bruce Banner into the Hulk. It always started in the eyes. You could see the storm coming. And not surprisingly, this cinematic mutation was also my favorite part of *Manimal*. Even more than with Banner and the Hulk, the special effects used to show Chase's on-screen shape-shifting were the real star of the show, and regularly got top billing, eating up minutes of every episode, keeping me and about nine other people in America rapt and transfixed as if we were witnessing not just a physical transformation but a spiritual transition as well.

I'd pitch forward in my chair, all the lights off, the glow of the Zenith casting the room in bluish tints; and my fingers curled tight over the arm of the chair, beads of sweat rising from my brow. I watched and witnessed, amazed and titillated anew each time as Chase mutated into a big black cat. I could not look away. . . . How could anyone?

The metamorphosis happens slowly, methodically, like a seduction, with that uniquely synthesized and sleazy porn-quality music playing in the background—because everything in the late '70s and early '80s had porn music in the background. But it feels right. Perfect, actually. Because you feel a little dirty for watching. Chase's shirt splits down the middle as his back swells and arches, his spine rising up in a thick, ropy ridge of knuckles. Guttural sounds burst from his mouth, as if it hurts to change, as if he's coming through something painful and ecstatic. His face bubbles and stretches, his muscles bulge, inflating like balloons. His jaw distends, his nose turning up and flattening out into a snout. Hair grows like weeds in a time-lapse video; and his fingers curl inward, melting into paws, as claws sprout from them. Fangs burst from his gums and his eyes change shape and color—all of it happening in slow motion and climaxing with the requisite stock sound footage of a panther's orgasmic growl.

And before you know it, it's over. Man becomes animal. And you're spent, done, exhausted, and satisfied. You don't even care what happens next. You just know it was good and you feel like you need a cigarette.

In other episodes, Chase also changes into a bear and a bull, a dolphin, horse, and even a snake, some of them happening off-screen, but all of the mutations under his control. All of it patient and careful. All of it thoughtful.

The title itself is so simple and so brilliant, an inelegant but efficient portmanteau. Man to animal. *Manimal.* How could it not succeed? How could it not be a huge success?

But in addition to competing with the super-popular nighttime soap opera *Dallas, Manimal* suffered from the story-telling challenge of finding increasingly novel situations in which changing into a jaguar proved narratively satisfying. Call it the seven-episode itch. Call it bad writing. Or call it the short, flaming brilliance of novelty, and the transient specialness of special effects.

The show was canceled after only eight episodes, and *Manimal* is today widely considered to be one of the all-time worst TV shows.

It maintains a cult following, perhaps because of its extreme badness and camp quality, perhaps because of its ridiculous premise and cheesy special effects, or perhaps because *Manimal* still speaks, much like *Beastmaster* and *Man from Atlantis*, to a deep-seated desire many of us hold to cross over, to live between two worlds, and to *master the secrets that divide man from animal, animal from man.*

That day at Disneyland, I watched the Beastmaster climb into his teacup for a ride and it was like seeing a piece of my childhood—a character who seemed to contain all of them, one man to represent all the barbarians of my youth—queued up for a ride; and I had the urge to run after him, to elbow tourists aside and leap into his cup because the experience was like watching that piece of my childhood, something that I realized I might never be able to adequately share with my own children, spin away from me, twirling off madly into oblivion, off into the place where all the '80s manimals still live, captured forever and contained in the illuminated museum of memory, never to return.

7.

THE INCREDIBLE HULK REMAINS, in my opinion, still the greatest, most psychologically and existentially compelling (not to mention commercially successful) protagonist of the conflict. He embodies something I'm circling around in this exploration—the savage and the sublime, two sides of the same human animal, beast and thinker, and the eternal struggle for balance and control. The Hulk is rage unchained—both protagonist and antagonist—and at least for a while, the star of a popular TV series that both captivated and terrified me as a child, a show I watched religiously until I was forced to stop by recurring nightmares of Lou Ferrigno.

In the comic books, the Hulk is barely even a hero, often playing the role of monster, a violent antagonist to other superheroes in fancy costumes. He'd show up in his tattered shorts and pretty regularly kick Iron Man's ass; in general, he wreaks some serious havoc wherever he goes, like a sentient storm rolling through an urban landscape. When Banner becomes the Hulk, when he crosses over, he loses control, loses his humanity; and in this way he is fundamentally different from Manimal or Hank McCoy, the Beast in the X-Men comics—a mutant who, despite his outward appearance as a big blue sasquatch or wolf-ape or something, maintains his humanity, control, and intelligence. He's articulate, well-spoken, and rational, even when he's beastly.

When Bruce Banner transforms into the Hulk, he transforms into chaos. He becomes a big green void, damaged and dangerous.

He's mute and dumb and violent. A monstrous force of nature. But when he is able to control his rage, when somebody isn't pulling on his chain or rattling his cage, he is the polar opposite of the Hulk; he becomes the gentle soul, scientist, and, in the TV show, the grieving husband who couldn't save his wife from a car accident.

The 2003 Ang Lee adaptation, *Hulk*, a visually stunning movie panned by critics and, not surprisingly, loved by me, is admittedly a film burdened by Eric Bana's acting (not to mention one of the most incongruous soundtracks ever recorded) and a script that, at times, becomes an obvious vehicle for studio animators to show off early CGI innovations. But despite its obvious flaws, the story still manages to capture the enduring conflict for Banner:

"Do you remember anything from when you were changed?"
"It was like a dream," Bruce says.
"Of what?" Betty asks.
"Rage, power, and freedom."

In this scene at the breakfast table in Betty Ross's cabin, Betty tells Bruce that she thinks his metamorphosis, catalyzed by "an accidental dose of gamma rays," is triggered by his anger, and that it may be connected to some deep emotional trauma. The Hulk, thus imagined, becomes a kind of metaphor for PTSD.

"You know what scares me the most?" Bruce says, "Is that when it happens, when it comes over me and I totally lose control . . .
I like it."

The rage is still there, still burning beneath the surface, still part of his nature. Bruce Banner, thinker and friend, loner and drifter, is not the absence of rage. He is rage and chaos restrained and balanced. Perhaps what we call intelligence is really just effectively restrained rage, just an extended suppression of our savage nature, and beneath every thinker and scientist and artist is an animal, a beast barely contained.

8.

WHEN I VISITED THE Bronx Zoo in September of 2014, after taking
a spin on the Bengali Express Monorail through Wild Asia, I wandered through the zoo's central plaza, past the abandoned monkey
house and the bronze rhinos, and made my way to the Congo Gorilla
Forest, where I watched mountain gorillas munch on stalks of celery.
The viewing area was entirely enclosed in glass, like a bubble, and surrounded on three sides by the thickly forested habitat.

The gorillas who weren't eating sat quietly, hands folded like
monks, staring into the nothingness of glass, where just beyond the
bubble's boundary, crowds of other less hairy, slightly more evolved
primates gathered, pointing, pushing, and holding their phones up to
the glass to take pictures.

I drifted away from the crowds surrounding the larger celery-eat-
ing gorillas and found a smaller gorilla, probably a female or a younger
male, sitting and watching the show. Her legs were tucked up to her
chest, her arms crossed, hands resting, eyes wide and impassive. I'm
tempted to call it sadness that I saw in her face, but I know that I'd be
projecting a human emotion onto the gorilla. Perhaps the look on her
face is closer to a look of what we might call resignation. Maybe it was
even something you could call peace I saw as I held up my phone to
the glass and snapped her portrait.

When I studied the picture a few moments later, I noticed,
suspended just above her left shoulder—like some kind of strange,

studio-produced awkward senior portrait—my own face and torso, a black cap shading my eyes. I seem to emerge, ghostlike in the scene, blending into the rock. The camera captured my reflection in the glass and superimposed it with the gorilla until the two of us, man and animal, blur together, inhabiting the same space in a way that we never could in reality. Through the common miracle of shutter and light, I became one with the gorilla, as if we were sitting together staring out at the strange pale faces beyond the glass.

FACING THE BEAR IN ME

1.

A CONFESSION: SOME DAYS I wish I could change into a grizzly bear and disappear into the deep woods. These are the days, amidst the challenges of everyday life as a divorced father, that I wish instead I could roam the tundra alone like my grizzly heroes; I'd go fishing, try to avoid physical confrontations with dominant alpha-male types, kill a few straggling elk calves, dine frequently on fresh salmon, raid the random cabin or campsite, or maybe even maul an occasional tourist for shits and giggles. I can't deny that I tend to romanticize the male grizzly's life in the wild. It's probably not healthy.

I wouldn't, however, do well as a bear in a zoo. Captivity frightens me—so much that I find myself planning for situations where I might be held in such a state. Just the thought of being caged gets my hackles up and makes me angry; and as Bruce Banner says in *The Incredible Hulk*, "You wouldn't like me when I'm angry."

Despite my anxiety, or perhaps because of it, my daughter and I enjoyed spending a lot of time at our local zoo, which has housed, among other constantly changing residents, a grizzly bear, a tiger, elephants, siamangs, tapirs, an alligator, giraffes, kangaroos, and a couple of orangutans, who seemed relatively peaceful and accepting of their confinement. The orangutans always looked stoned and happy, more amused at the antics of human observers than anything else.

Recently, a young man climbed over a short fence at the Chaffee Zoo and approached the wire fencing where the orangutans sometimes hang out. Initial reports I read said he'd climbed into their cage, but I knew this would be impossible, given that their enclosure is covered on all sides with thick wire fencing. I'd checked. I'd looked for entry points. Later reports clarified that the man had climbed over a viewing barrier (really just a split-log fence) and reached into the cage, trying to touch one of the primates. He tried to touch a different version of himself. But he'd only succeeded in getting scratched by an orangutan and arrested by a police officer.

I read this story and remembered what I'd learned in a Physical Anthropology class in college—that humans are related evolutionarily to the individualistic orangutan as well as to the social chimpanzee. We're a mix of both: individualist apes that crave contact with others, and social apes that crave solitude. We are genetically, biologically, conflicted.

My daughter asks me pretty regularly to name my favorite animal. I love the tigers and the orangutans, the elephants and the sea lions, but my answer never changes. I always say, "Bear."

Bears have a power, grace, and dignity that I admire. Even their savagery is attractive to me. Bears, too, are pulled between their individualist and social natures. But I also know that male bears are the ultimate absentee fathers, the alpha assholes who have children with numerous women and never take any responsibility for them. They're the rogue gangsters of the animal world, which has never really been my style.

Worse than that, male grizzlies are known to prey on cubs, forcing the mothers to protect their young; and the truth is that, most days, I'm more like a mama grizzly, more sow than boar. I'm the one waking my cubs in the morning, herding them off to school, feeding them, dragging them with me to get food, teaching them how to survive, and trying to protect them from harm.

2.

I ONCE DESCRIBED MYSELF to my therapist as feeling like one of those trained dancing bears, tugging against a chain, or like one of those bears they used to chain up in a ring, pitted against a bull in a fight to the death. I told her that I've never seen myself as the bull.

"What's the chain?" she probably asked. "What's making you dance?"

"I don't know. Family, marriage, fatherhood, work, money . . . life."

I told my therapist I was afraid sometimes that the stress would get the best of me and I'd break the chains, afraid that I would finally lose it at a meeting or in traffic, hopefully not in front of my kids, and I'd explode. Burst. Erupt. And disappear. These are the words we didn't want to use. I am a large man. When I explode it could be messy and loud.

Flooded is the word we used in therapy: "I feel flooded," I would say to myself because it was important to recognize when I was beginning to lose control of my emotions and feeling stressed out.

It happened fast. I didn't hurt people physically. But I yelled and cursed, and I could hear my voice get big, like it inflated and barreled out of my chest, almost with a life of its own. It could be scary. My angst came out in a primal roar. Curse words spewed forth as if I was speaking in tongues. I didn't want to hurt or frighten anyone, and it was mostly small things. Objects. Lost keys, computer problems, spills, the tiny frustrations of everyday life that, if I wasn't careful, could sometimes pull the trigger or break the dam. Mostly I yelled at myself.

But it did feel like I was overflowing, like all this stuff was pouring out and I couldn't plug the holes fast enough.

Or it was like a switch got flipped and I'd want to smash things, want to rage and storm and howl.

"Count to ten," my therapist said.

But I couldn't even find the numbers sometimes.

I'm better now. Therapy and time, happiness and calm have prevailed, but sometimes I still get stressed or tired and I feel flooded. Some days I still feel like a bear on a chain or the Hulk, barely restrained. I've learned to control it better and to apologize when I get flooded.

My therapist has helped me feel normal. We talked a lot about balance and authenticity. Or I talked. And she just asked me a few questions.

We talked calmly about my angst and rage.

We practiced flood control.

3.

In HIGH SCHOOL I was never in theater, never acted in a school production, unless you count my performances on the basketball court. I have a terrible singing voice and I couldn't dance, but I could act on the court, and I could talk. I could perform like I was supposed to be there, even if I was outmatched, out-quicked, and outsized most of the time. I had no choice, really, but to step up and act the part. My performance was a matter of survival. Or at least it felt that way to me.

At 6'4" in shoes, listed an inch taller in the program, I was the shortest starting center for the biggest high school in the state of Kansas. On defense, I regularly had to guard post players who had six or seven inches and twenty pounds on me—or I had to guard 6'6" wing players who were thinner and quicker and could jump out of the gym if I let them get past me.

I am not fast. Not quick. I had good footwork, which only meant that I knew where to put my body to create the best angles for easy shots or to impede the progress of my opponent. I had a good shooting touch and a pretty sweet turnaround jumper. But my greatest skill on defense, and probably what got me the most playing time, was that I was a beast in the paint—a slapping, pushing, elbowing, shit-talking thug who would knock the crap out of you if you came in the painted area beneath the basket. Admittedly, much of this was a calculated performance that, most days, seemed very far from who I was then or who I am now. But at my worst (or best), I could summon the flood of rage and bring it down furiously upon my opponent.

At my worst on the court, I became a savage barely contained, hardly harnessed by the rules and the refs. Once, in the heat of the moment on the court's stage—a community center, after high school, in my Kansas hometown—I told another player that I would "rip his head off and shit down his neck," the logistics of which are mind-bogglingly grotesque and ridiculous to the point of absurdity; but, perhaps not surprisingly, he kept his distance from me for the rest of the game. I guess I'd gotten in his head—so to speak.

The truth is also that, at my best on the court, I felt as infinite as the Incredible Hulk—big, angry, and unstoppable—and, at my best, I was able to channel a kind of unspoken existential rage into something bordering on grace and skill. Some of my best games were during City League play in the years after my brother was killed in an auto-mobile accident, when I played with a reckless abandon and rage that, unfortunately, occasionally boiled over on the court. I was, at times, nearly unstoppable and unflappable, a force of long hair and muscle and noise, until the role itself became like a drug.

When it comes over me and I totally lose control . . . I like it.

In high school, I wasn't a star, wasn't the best player on my team, and I knew part of my role was as a kind of enforcer, a basketball goon. I was proud when people told me they liked playing with me and *hated* playing against me, proud when the parents and family of my oppo-nents turned against me, screaming at me from the stands, as if I was the orchestrator of their unhappiness and angst. It was probably too much power to give a seventeen-year-old kid.

4.

I QUIT PLAYING BASKETBALL years ago and started playing racquet-ball again just before my daughter was born. I continued playing rac-quetball through the hard years of my divorce a few years later, into my late thirties and early forties, before I reinjured my knee and my doctor told me the only solution was a new knee, an artificial one.

"Can I get a bionic knee?" I asked. "You know, and like leap over walls and stuff?"

My doctor has the demeanor of a tree stump, so I didn't expect him to laugh. And he didn't. Perhaps he understood that jokes were my way of deflecting, that I was trying hard to make light of the fact that I'd lost something. I'd lost my knee (there's literally nothing holding it together except muscle and skin) but, more importantly, I'd lost the sanctioned release of aggression and angst that the racquetball court provided. I'd lost my rage outlet.

I guess I needed to smash a ball against a wall, needed the cha-otic noise and movement, and the chemical rush of endorphins, more than I realized. My therapist assured me such release was good for me, good to have that outlet for my angst. Plus it was a lot cheaper than my couch sessions.

Afterward, I'd come home happier, peaceful, and ready to juggle it all. I know it sounds so stereotypically male and I suppose it is. But it was better than blowing up and yelling over stupid stuff or breaking down in tears at iPhone commercials.

I'd also seen, however, the dark side of this sanction, this permission that sports allow.

The second varsity game of my junior year in high school we played Shawnee Mission South, a team that featured not one but *three* future Division One players.

I was matched up that night against South's 6'10" center, Dan Augulis, a name that, in the days leading up to the game, had begun to take on a kind of mythic and monstrous presence in my mind. Much of the focus in practice and in the halls of school, even from the scant local media, was how I, at 6'4" and 185 pounds, was going to possibly stop Dan Augulis.

People wondered if I could do anything at all to keep Dan Augulis from dominating the paint. After all, I was definitely NOT a D1 prospect, and Dan Augulis had a legitimate six or seven inches of height on me. Dan Augulis began to seem monstrous and mythic, bigger than reality.

Dan Augulis

Dan Augulis

Dan Augulis

My dad had always told me that if you could get your opponent to focus on you, get them angry or frustrated with you rather than focusing on the game, you'd already won half the battle. I figured I'd already lost half the battle to Dan Augulis before the game even started, so I had a lot of making up to do.

From the minute the ball was thrown up for the tip-off, one that Dan Augulis won easily, my sole focus was on psychologically breaking him. I didn't just want to intimidate Dan Augulis. I wanted to humiliate him. I wanted to destroy him. My unexpected role as intimidator might have been helped by the fact that, while playing a quarter in the Junior Varsity game earlier as a kind of warm-up, I'd taken an elbow to the eye that split my brow open and sent blood pouring down my face. The trainer had to tape up my brow with a butterfly bandage and, by the time the varsity game started, my eye had swollen up purple and half-closed. I looked like a boxer who'd taken a beating already and was just crazy enough to be itching for more rounds with the champ.

Dan Augulis was tall. Really, really tall. When he caught the ball and I extended my left arm up to defend him, my hand only reached up barely past his elbow. He could shoot over me any time he wanted; so when he caught the ball and turned to face the basket, I'd reach up with my left arm, waving it in his face.

Then I'd attack.

I'd cock my right arm, clenching my fist and curling it up to my chest, transferring all the power and leverage to my shoulder. As Augulis raised the ball up to shoot, I ducked in under his arms and delivered a quick hard elbow strike to his rib cage.

Pow.

And another one. *Pow.*

The first time I hit Dan Augulis in the ribs, I could tell he wasn't sure what to make of it, how to deal with me. The second time I hit him, he looked scared. I watched him watching me instead of his teammates. And I knew it was over. I had him. Every time he caught the ball, he physically recoiled from me, curled away, and backed off.

He was afraid of me. He was afraid of getting hit.

And I *fed* off of this fear. I swelled with it, charged as if I'd been plugged into an outlet.

I called Dan Augulis bad names and told him that he wasn't as good as everyone said he was, that he was no D1 prospect. I mocked every missed shot and laughed at him. It didn't matter that I wasn't scoring much because he wasn't either. He missed a wide-open dunk on a fast break and I howled at him, telling him how pathetic I thought he was.

We kept it close the whole way and only lost by a few points. It was one of our best games of the season against one of the best teams in the state. But it was also perhaps one of my worst moments as a human being. I still feel guilty for it, and still remember his name when I've forgotten hundreds of others I played against. Sure, part of my metamorphosis that night was because I'd allowed everyone else to intimidate me, to get in my head, and the way his name had haunted me—Dan Augulis, Dan Augulis—keeping me awake at night. Part of it was a performance of success and survival, a feeling of being backed into a corner that I knew I couldn't sustain. Part of it is simply my competitive nature.

I kept playing city league basketball after high school in most places where I lived, but eventually gave it up in part because my knees just couldn't take it any longer, but also in part because I realized that the court often brought out the worst of my competitive side, that sports, even the game I loved dearly, legitimized a transformation that I didn't welcome.

I became a different sort of person, a man I didn't recognize—or didn't want to recognize. It was a performance of self as predator and aggressor that I wanted to quit. I'd become a beast on the court because I had to in order to be successful, but I didn't like the way the change haunted me. It wasn't the real me or even a character I wanted to play any longer. It was a part of me I had to let go, or at least try to—even if I understood that it would always be there, hiding beneath the surface.

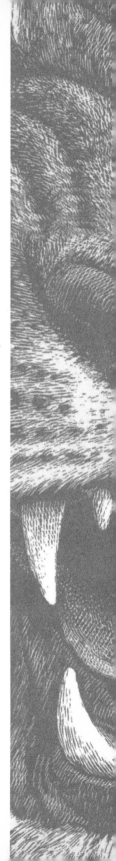

PART FOUR

THE
ANIMAL
WITHIN

FIGHT, BULL

We are all ready to be savage in some cause. The difference between a good man and a bad one is the choice of the cause.

−WILLIAM JAMES

VESSEL OF VIOLENCE

A friend of mine, a long-haired, bearded carpenter, told me a story about a time he boarded a city bus in Fresno and nearly resorted to violence—the kind of violence they write stories about in the daily paper, a different sort of attack narrative, a different kind of cage. *Your memory isn't even working normally.* It could happen to you or to me. It could happen any day.

The carpenter remembers climbing the steps into the air-conditioned bus, his eyes quickly scanning, surveying the space as he walked the narrow path to an empty seat.

The army had taught him to see, to be aware of danger, but I'm not sure he expected this. An overweight girl followed behind him, nobody he knew. Just a girl. A big girl he'd seen waiting at the bus stop. The two of them moved carefully, close together through the door.

If you didn't know you might think they were a couple. And before either of them could find a seat, a noise rose up from the back of the bus, an unmistakable ruckus of sound. The carpenter couldn't miss it. Others must have noticed too. Three boys began to hoot and laugh, mocking the girl, mooing like a cow at her and slapping their thighs in hysterics. *In a situation like this, in your mind, you're in a combat situation. Your mind is functioning. You're not thinking in a normal way.*

The carpenter found a seat near the girl. She tried to ignore them, but they kept going, kept mooing like cattle. They weren't happy with simple humiliation. This carpenter boiled and watched and listened to them, trying to tamp down his anger. It was wrong, what they were doing. The girl took their abuse for a few blocks, took it with her chin down, tucked to her chest. Eventually she moved to another open seat near the front of the bus. *You are so hyped up. Your vision actually changes.*

It's not long after this that the teens started in on my friend, the carpenter. At first he didn't understand, didn't realize they were calling him "Geico," and saying, "Hey, Geico? Geico?" in reference to the insurance commercials featuring hairy-faced cavemen. But eventually he, too, understood that they were mocking him for his appearance. He wore a full beard and long hair, sometimes a vest and blue jeans. He wore functional glasses with black frames.

Perhaps because there were three of them, they felt safer, stronger, more brazen. Perhaps they baited him because they hadn't seen this bearded carpenter, in an effort to defend his friend, grab a belligerent drunk at a party, slam him against a wall, and throw him to the ground; hadn't seen this carpenter deliver a swift violent kick to the drunk's head, and certainly hadn't held him back from kicking again, caving in the drunk's skull. Would he have done it? I don't know. But I saw in his eyes that he could do it if he wanted to. When I'd held him back, my arm across his broad chest, my mouth whispering into his ear, "Easy, brother. Easy," I also knew that I was holding myself back too. How quickly we can cross over. Prey to predator. Man to animal. Civilized to savage.

Your field of view changes. Your capabilities change. What you are capable of changes. You are under adrenaline, a drug called adrenaline. And you respond very quickly, and you think very quickly. They mocked him because they could, because this man, this bearded carpenter was too smart to bite, because he had already turned the other cheek. He ignored them. Or tried to. *That's all. You think! You think, you analyze, and you act. And in any situation, you just have to think more quickly than your opposition. That's all. You know. Speed is very important.*

Even in that moment when the encounter went from offensive to threatening, as one boy came and sat down across from the carpenter, staring openly, daring him, taunting him openly, he didn't take the bait. The boy leaned over the aisle, cooing, "Geico, Geico. Geico," close enough to smell, and as he twisted his face close, trying to catch the eyes, he said, "Geico?" again like a test, right in this bearded carpenter's face.

This better man didn't turn animal there on the bus, didn't smash the boy's young face, didn't beat his head against the floor or one of those metal poles until it split open and blood bloomed from the crack. *My intention was to murder them, to hurt them, to make them suffer as much as possible.* He didn't produce a gun suddenly there in his hand, didn't shoot all of them, didn't put a hole in the last boy's back, a clean round hole like a birthmark. *If I had more bullets, I would have shot 'em all again and again. My problem was I ran out of bullets.*

The carpenter didn't do any of these things because we have long since left his story and drifted into a story shaped by Bernhard H. Goetz and his gun, his fear, racism, and rage, and by a culture that embraced the expectation that violence is the easiest answer. In 1984, the year I turned thirteen, Bernie Goetz took a gun onto the Number 2 subway train in Brooklyn and shot four unarmed black men who he claimed had threatened him with a screwdriver. He was later acquitted of all charges except possession of an illegal firearm. *I was gonna, I was gonna gouge one of the guy's eyes out with my keys afterward.*[6]

---

6. Author's Note: All italicized passages in the first section are quotes attributed to

This bearded carpenter didn't go after the eyes, didn't even tell me that he wanted to. Like me, he doesn't own a gun, certainly wouldn't carry one onto a city bus, and doesn't seem capable of such melodramatic violence; perhaps it's because this man, though he, too, grew up with Goetz, is no vigilante or stock character of the eighties, and he knows how to control his violent capabilities. He knows his limits and switches and how to lock them down. His story thus became a vessel for my own imagined defense, my own burdens of fear. I remember this carpenter's story as both a warning and a lesson. I try to open it up again here on the page, to see again how he resisted the urge, how in the face of a threat, he turned inward, looked away, and waited for the bell to signal his stop.

Bernie Goetz, taken from the following three sources: "*Your memory isn't . . .*" to "*Speed is very important . . .*" from Myra Friedman, "My Neighbor Bernie Goetz," *New York*, February 18, 1985; "*My intention was to murder . . .*" from "'You Have to Think in a Cold-blooded Way'" [transcript of Bernhard H. Goetz police interview], *New York Times*, April 30, 1987; and "*If I had more bullets, I would have . . .*" to "*I was gonna, I was gonna gouge . . .*" from Margot Hornblower, "Intended to Gouge Eye of Teen, Goetz Tape Says; 'My Problem Was I Ran Out of Bullets,'" *Washington Post*, May 14, 1987.

PRETEND TOUGH

I'm quite sure my son has never stood before a mirror as Robert De Niro does in *Taxi Driver*—stripped to the waist, lip curled in a snarl, arms flexed—and practiced intimidating facial expressions and quips.

You fucking looking at me?

I know he's never said things that Clint Eastwood says in *Dirty Harry*, lines that Charles Bronson delivers in *Death Wish*, or catch phrases from Arnold Schwarzenegger, Sylvester Stallone, Russell Crowe, or someone else.

You feel lucky, punk?

Perhaps he's never imagined or pretended how he would fight his way out of a tough situation, never dreamed of movie-quality, *Gladiator*-style vengeance against the men who harmed him or his wife or mother or sister or, worst of all, his children.

I like to think he's a very different sort of boy than I was. We've made an effort to shield our son from violent movies and video games, to encourage him to always treat others with kindness; and, for whatever reason, he doesn't seem to have a predilection toward these stories in the same way that I did. Then again, he's not growing up in the late '70s and early '80s.

My identification with action heroes and violent men seemed a normal response to the times. I've had very few opportunities to ever put the lines into practice, but I've spoken those of Al Pacino and De Niro and Bronson—or I've imagined myself speaking them—and I've internalized stories focused on the violent unhinging of one man's world, and I've imagined myself capable of movie-quality barbaric violence, pictured myself doing terrible things to people as a way to protect myself or my family.

In the original *Death Wish*, Paul Kersey's wife is beaten and his daughter raped and brutalized by a gang of thugs who follow them home from a supermarket. His wife dies and his daughter is left in a catatonic state.

It's an awful scene. Devastated by the attack, Kersey, played perfectly by the painfully wooden Charles Bronson, takes a business trip to Tombstone, Arizona, where he visits the famous OK Corral. As if some seed of this old cowboy story takes root, an idea sprouts in Kersey's head—an idea clouded with an anger to which many people in America could apparently relate. He returns to New York, bent on vengeance, and proceeds to systematically hunt down people he considers of the same general kind as those who attacked his family, providing Bronson with opportunities to deliver his own brand of Hollywood vigilante cowboy violence—a brand that just happens to be pretty damn racist. Part of me, I realize now, was always rooting for the bad guys in these movies.

In one particularly eerie echo of the Bernie Goetz story, Kersey buys a bag of groceries that he doesn't need and carries them onto a subway train. He sits there like a spider lurking on his web, the bag of food dangled into the aisle like bait, and when a black boy bites, grabbing for the groceries, Kersey shoots him dead.

If you're susceptible to such films (and I apparently had a predisposition to them), you find yourself emboldened, even empowered by these stories and troubled by the violence, the fear and anger, the distrust and racism that lurks within many of them. Eddie Murphy used to talk about the "*Rocky* effect" in his stand-up act, that indestructible, invincible feeling that Italian guys get after watching *Rocky*, as if empowered by the film, they, too, are capable of going toe-to-toe with a real-life Apollo Creed or Clubber Lang. Psychological research into something called the "Script Theory of Aggression" suggests that children learn violence and aggression from observing scripts and mimicking the roles.

I admit that it's hard for me to accept that our thoughts and behavior might be so heavily and perhaps negatively influenced by the stories we see and hear; but at the same time, as a writer, I cling to the belief that words and books can also have a lasting *positive* influence on readers. Right or wrong or just a complicated reality, books, television, and movies—and the stories we consume through them—begin to shape our interactions with the world and our responses to aggression and violence; and it seems that real trouble often comes when our imagined defenses, shaped as they are by violent fantasy, slam up against a true threat.

SMELL THE FLOWERS

The Story of Ferdinand by Munro Leaf is a children's classic with black ink renderings of a young bull that is not like the other bulls. While the others are out in the fields butting their heads and stomping their feet and acting, in general, like young bulls, Ferdinand retreats to the shade of his favorite cork tree, where he prefers to "just sit quietly and smell the flowers."

Despite his individuality or perhaps because of it, Ferdinand grows to be the biggest and strongest bull of the bunch. Soon the men in funny hats come to find the fiercest bull for the bullfights in Madrid. Ferdinand, uninterested in the men or in fighting and being fierce, returns to his favorite tree for some rest and relaxation. But instead of sitting on the grass he sits on a bee.

The bee, being a bee, stings him and Ferdinand leaps into the air, snorting and pawing the ground. Because of this physical display, Ferdinand is mistaken for an aggressive, fierce bull and shipped off to Madrid in a rickety wooden cart. He's thrust into the ring, named "Ferdinand the Fierce," and surrounded by fans clamoring for blood. The tension builds around the pomp and circumstance of bullfighting as the *banderilleros, picadors,* and the matador all parade into the ring. Ferdinand finally makes his entrance, trots out to the center of the bullring, plops down, and promptly refuses to fight.

He sits there just quietly and smells the flowers in all the lovely ladies' hair. He is unquestionably heroic. The matador, who can't show off with his sword, is shamed to tears, and Ferdinand is shipped back home to sit under his favorite cork tree.

Published in 1936 during the Spanish Civil War, the book has often been called a "political" text and Ferdinand considered a "pacifist" character. The book was banned in Spain, burned as democratic propaganda in Germany, and even regarded in the United States as being detrimental to its efforts during World War II.

Ferdinand, it seems, was one of the first hippies. In the face of ritualized competition, cultural homogeneity, and violent aggression, he resisted. When misjudged, pushed into the ring, prodded and provoked to fight, he chose the path of nonviolence. When expected to be fierce, he stopped to smell the flowers. He would've been right at home in California during the Summer of Love. But if Ferdinand had happened to find his way to the West Coast of North America during the Gold Rush, he might've faced a very different opponent in the ring—the now-extinct *Ursus arctos californicus*, or the California Grizzly, the Golden Bear.

One popular form of entertainment and spectacle at the time were orchestrated fights between Spanish bulls and wild California grizzly bears. Often these fights took place in a central town square or outside of the local Spanish missions or on the grounds of local Mexican ranchos; and they were regularly held on Sunday, after church services ended, when parishioners would gather in the town square for some afternoon entertainment—yet another Church-sanctioned form of ritual and worship.

The bull and bear were roped together in the ring, then prodded and antagonized with long wooden poles or spears. Often the bear was chained to the ground, preventing his escape; and neither of them would've been allowed just to sit quietly and smell the flowers.

They were forced to fight; and the resulting battles were especially violent and gruesome fights to the death, often (but not always) won by the bear. This uniquely Californian form of bloodsport pitted the civilized, domesticated Spanish bull against the savage and uncivilized grizzly bear; and the people gathered to witness the carnage, to revel in what one eyewitness saw as California's collective, "squalor, misery, and shame."

But these stories are more than just evidence of the savagery and "rottenness" that ruled California; they also become, for me, parables of westward expansion, capturing in stark relief the age-old conflict between the misunderstood savage, the retreating wild, and the untamed pitted violently against the forward charge of civilization and domesticity—a conflict embodied by Ferdinand and Treadwell, Tyson, Haas, and Villalobos, and all the other charismatic barbarians. It's an old story, an ancient story—bear vs. bull, Roman slave vs. lion, gladiator vs. gladiator, boxer vs. boxer—that rarely ends peacefully in a pasture, sitting quietly beneath a cork tree. Instead it ends in a pit, a ring, a stadium, at a party with animals watching animals destroy each other.

REMEMBERING THE DANCE

The summer after we moved to Fresno I was hired to teach for a writing program in Madrid, Spain. My wife was three months' pregnant at the time, and we traveled there with our five-year-old son. As part of an "excursion" through the writing program, I had an opportunity to see a bullfight in the central bullring of Madrid, the very bullring where Ferdinand had famously and fictionally refused to fight.

One year before I visited, an experienced matador was gored in his groin. The bull's horn ripped through the matador's pants, pierced his scrotum, and liberated one of his testicles. An image in the newspapers the next day showed the severed nut flying through the air, suspended above the dirt, with the blurry crowd cheering in the background.

This seemed like an opportunity I couldn't pass up.

I did, however, decide *not* to bring my five-year-old son along for the show. I'd heard that while full of pageantry and history, ritual and even beauty, bullfights are also incredibly brutal, bloody, and violent. I'd heard that sometimes the matador loses a testicle. I wasn't sure that was a script I wanted my son to internalize.

The fights, or *corridas* (the translation is closer to "dance" than "fight"), featured three matadors—two men and one woman—versus three different bulls. It took me a while to learn the rituals, but I eventually picked up most of it. Whistling is booing. Waving a white hankie is cheering, fawning even. And if an old Spanish woman offers you a pull off of her bota bag full of cheap red wine, do NOT, under any circumstances, put your lips on the nozzle or she will smack your knee and curse at you.

There were other things to learn: if after the fight the matador places his hat in the middle of the ring, it means something significant. If he places it upside down, it means something else. Hemingway would know, but I don't remember for sure what these things meant; at the time they seemed really important. I spent a lot of time focusing on the details, the small things, and the tiny rituals of sport. I did this because there were many things about that day that I didn't want to remember.

I didn't want to remember the blood. And the preening—matadors posing and strutting, thrusting and prancing, turning their backs on the bull as a show of disrespect, a disregard for the bull as a threat, a way to say with their bodies, "I am not afraid."

I thought to myself, *They look just like a smack-talking basketball player or a touchdown-dancing football player. Or like a very large rooster, brightly colored, strutting around a dirt yard.*

I didn't want to remember that the bull refused to die easy, that he was nothing but fight and resistance, nothing but hard animal defiance. The bull never quit trying to survive. I'd never seen an animal fight so hard to live.

I didn't want to remember the matador, failing to kill the bull with his sword, his pink pants festooned with gold filigree, soaked in deep wine-red blood, and how he jabbed a three-pronged sword into the bull's neck again and again, working hard to sever the spinal cord.

I didn't want to remember the bull staggering, his head lowering, neck muscles compromised, and his horns wagging back and forth. I was hoping that, in one final gasp, he might lunge forward and liberate the rooster-man from one of *his* tiny testicles.

I didn't want to remember how, only when the bull was crippled and weakened, could the rooster-man sink his sword into the bull's neck, severing the spinal column and dropping the bull first to his front legs, where it paused for a second, before crumpling over dead.

I'd never seen so much blood.

I didn't want to remember the way it spurted out of the bull, and sprayed all over the rooster man. This is not hyperbole. There was so much blood. It gushed out of the animal, pressurized as it was by the bull's pumping dying heart. It is impossible for me to describe how much blood. Whatever you imagine, multiply it by ten, by twenty, by one hundred. Fountains of blood. The dirt, after centuries of such slaughter, must have been soaked with it, the soil itself made as much from blood and death as anything else.

The only thing I really wanted to remember is how I loved the female matador deeply and profoundly for her killing skills. She killed the animal fast, with little blood and gore, little suffering. She was so much better than the men at the killing game. One swift strike, one plunge of her sword, and the bull died mercifully. I loved her for this precision, still love her for this because the truth of it rubs against the grain of stereotype. She alone possessed the instincts and talent to give the bull a somewhat dignified death . . . if such a thing actually exists. Probably it doesn't, at least not for the dying. But she seemed to offer the animal the sort of respect that I thought it deserved. The crowd agreed and, after her first kill, she was given one of the bull's ears as a trophy.

I understood finally that my friend Ferdinand would have been prodded to fight, then killed, slaughtered for sport and ritual in the ring. I knew that once you are thrown into that kind of cage, you have no choice but to play the role you've been assigned. Ferdinand would have been killed, butchered, and sold in the abbbatoir beneath the stadium. This is how the dance goes.

If Ferdinand had been so lucky as to have a quick death, the matador might have been gifted one of his ears, cut freshly from his head; and if the *corrida* had been particularly dramatic and compelling, if Ferdinand had somehow been goaded into fighting especially hard and theatrically before he died, the matador might have been given the bull's tail as a trophy.

FIRST-DAY PLANS

Mike Tyson once said, "Everyone has a plan until they get punched in the face." I have a plan for even the most ridiculous threat situations. When we were in Alaska, I had a grizzly bear attack plan. Walking with my kids in Fresno, I have a pit-bull attack plan. I even have a "First Day in Prison" plan. It has different variations but, unlike the Ferdinand scenario, my plan doesn't rely on passive resistance. No, it involves sending a message to other inmates through unnecessary and extremely barbaric violence. It involves compound fractures, crushed windpipes, and other really gruesome stuff.

There's no good reason why I have a plan for my first day in prison except that I've seen too many movies; but I don't think I'm alone in my tendency to mentally script out a response to such seemingly unrealistic threats. I am not a violent man. Mostly I abhor it and have avoided it at all costs. But that doesn't mean I don't think about it.

I think many people do this. At any given moment, we might project ourselves into an imagined near future and see our new self acting and reacting to a variety of threats and dangers. This is what allows us to drive a car in traffic without freezing up. And such quick, dynamic, and imaginative thinking is at the heart of the dilemma I'm trying to explore. This acting, this fear-driven role-playing, is a desperate sort of practice for the violence that many of us are raised to believe will inevitably befall us. And it encourages us to live in between the present and the potential future, a kind of liminal space wherein all possibilities are a matter of statistics.

Our loved ones will be threatened or killed. Trusted institutions will fail to protect us and may even victimize us. We will be attacked. Our friends will turn on us. It's just a matter of time. We will be jumped, mugged, punked, or sucker-punched in a bar. Animal or man will attack us. Or maybe we'll be targeted at a party, one of those end-of-the-year parties where everyone is drunk and loud and laughing, when everyone is letting out a lot of stuff. No matter how hard we try not to, how innocent we might be, we know we'll be pushed into the cage of an old story, forced to play a role that quickly transcends our imagined practice.

BEAR OFF HIS CHAIN

This is how your story goes: It's spring and things are blooming in the Central Valley of California. You have a night away from parenting. Your marriage swings in the balance. You're trying to be a teacher, a father, a professor, and a writer—trying to juggle all these roles. You feel ripened, soft, and ready to split open. And you're sitting at a table talking with a group of friends and colleagues at a backyard party. It's the end of your first year in a tenure-track teaching position and, more than ever, you feel like a trained bear tugging at your chains. You're just having a few drinks at this party, not standing out, not being loud or obnoxious, trying not to make yourself a target. And then from nowhere it seems, the husband of a student punches you in the left shoulder. You see his swing in your peripheral vision and barely have time to flinch. He punches you hard. Then he laughs. He's drunk.

He says, "Let's play the Fight Club game."

You want to tell him that this isn't that kind of party. Instead you say, "No. I don't want to play the Fight Club game."

You look up at him. "Please don't hit me," you say. You'd rather be like Ferdinand smelling the flowers. "Smell the flowers," you want to say to this man.

But no . . .

He hits you again. Pokes the bear.

"Seriously," you say. "I don't want to play. Don't hit me."

Smell the flowers, you tell yourself. *Smell the flowers.*

He's laughing now.

Someone else says, "Don't. Don't do it."

But he doesn't stop.

Instead he swats you open-handed on the back of the head.

Like a punk.

Like he's cuffing a mutt.

Like you're his dog.

And this is what does it. This is all it takes for you to split open, cross over, and break the chain. You can feel the darkness spilling out. You can feel the savage flood rising up from the inside. You can feel yourself crossing over to the animal side. He punks you in front of all these people, your colleagues and students, and he makes you look the fool. But he also makes you feel weak and vulnerable. You're old enough, smart enough to know that you should turn away. You're the Professor, after all. You should be the bigger man by refusing this lowly role, but you just can't do it. You can't stop it. You are inevitably and violently human.

You wouldn't like me when I'm angry.

You feel the tide turn, as if his dither knocked something loose and rattled a side of yourself that you keep caged up. You're six-foot-four and you weigh 260 pounds. You have a sizable scar on your face and a protruding brow. Some people think you look intimidating, and you've sometimes relied on this image to keep you out of confrontations. But you've never really been in a fistfight in your life. You've been able to bluff your way out of most situations. Your father always told you, "You never start a fight. But you finish it," and you took this to heart.

You tried to prevent this one. You tried to warn him not to hit you. You don't like to be hit.

You wouldn't like me when I'm angry.

You didn't jump into his cage. He jumped into yours. And though you didn't know for sure what you would do until he hit you, when he smacks you in the head, you know instantly and immediately that you will finish it, you will hurt him, and you will make it abundantly clear that he does not want to play this game with you. And almost as quickly, you also know *how* the violence will happen. The whole scene runs through your brain. You rehearse your lines and moves. Time seems to warp and bend to the story. You're playing a role you've rehearsed in front of a mirror.

You stand up slowly and push in your chair.

You know there is menace in these actions, know that your calm will make you seem even more frightening. You don't say a word as you turn to him.

And at this point he's backing up, away from you, grinning stupidly. But it is too late. He's crossed a line. Or you have. Perhaps an invisible line between reality and fantasy, between savage and human—a line that honestly worries you. But you'd made it clear. You'd asked nicely. You told him you didn't want to play his movie game. You rejected the role he cast you in.

Everyone watches and listens like an audience. You step up to him and grab him around the throat with your left hand. He's smaller than you, but not a tiny man. In one big push and heave, you slam him against the outside wall of the house. His head thuds dully on the wood siding.

With your left hand still on his throat, lifting up ever so slightly against his jawbone, pressing him into the wall, you point your right index finger into his face and say, very slowly, "Don't. Fucking. Hit me," and then you let him drop and you walk away.

ENCORE PERFORMANCE

When the show ended, I stepped away from that version of myself and quickly realized that it was one of the most intense physical confrontations I'd ever experienced, and something I never wanted to go through again. The adrenaline flush left me tired and shaky. I wanted to vomit. And cry. And then die. It was terribly un-cinematic. There was nothing sublime or ecstatic about it. I did not feel more fully alive in those moments. There was no spiritual transformation. At first I was afraid I'd seriously injured the guy. I'd clearly overreacted to what wasn't a real threat, clearly allowed him to push all my buttons and flip my switches.

Later, and for months afterward, I would be plagued with guilt and regret over the incident, replaying it in my head and trying to picture myself just walking away, just turning away from him and leaving the party. I thought there was a pretty good chance I might get fired or at least reprimanded by someone. Everyone saw it. I thought people would be afraid to be around me, thought my colleagues would regret hiring me and send me back to Kansas or Colorado or anywhere else.

Despite many people telling me things like "He had it coming," I knew this wasn't really true. Not really. I knew that as much as I didn't deserve to be hit and smacked, to be humiliated like that, he also didn't deserve to be slammed against a wall.

The whole thing was so stupid. It was as if the two of us had been sucked into a clichéd pop-culture narrative of masculinity and violence, of intimacy and savagery, and neither of us knew how to get out.

So we played our roles until the lines were exhausted, the audience confused, and everything collapsed, until it was no longer a game or a movie and became *too* real.

Here is what I know now: There was no redemption through violence, no true empowerment. There was no ecstatic awakening, no transcendence. At least for me. Maybe he felt differently. The movies lie perfectly. Even the facts hide the truth much of the time. The stories are delicious fictions. There was no hug between me and the guy at the party, no burying of the hatchet, no breaking of bread or some other Hollywood ending. Nobody cheered for me, or chased me out of the party, or slapped me on the back, or raised my hand in the air. Mostly everyone just seemed confused. There was only a smudge on the siding, a mark from his head, and the bruise on his throat. There was only my deep and abiding guilt and the fear that there is—perhaps in each of us—a part that is susceptible not just to violence but, perhaps more dangerously, to the self-aggrandizing narratives of violence as protection, salvation, and redemption.

I wonder if I still cling to the stories of my carpenter friend and Ferdinand the bull because they offer two extremes, two possible responses to the threat of physical violence, two examples of the human predator in action, and because they cause me to question: Am I like Ferdinand, the gentle bull who refuses to fight, preferring instead to sit just quietly and smell the flowers, or am I the sort of savage who can be provoked into a violent attack?

Yes. Yes I am.

PART FIVE

IRON
MIKE

SAVAGE HERO

*I have to reign. I have to reign tyrannical. I have
to be the beast. I have to be the savage. You know
what I mean, just to put in place the erudites and
the bourgeoisie and the people of this society that
believe that I'm trash and scum. I'll be trash and
I'll be scum, but I'll be angelic trash and scum.*

—MIKE TYSON

ROUND 1

. . . he has the power to galvanize crowds as if awakening in
them the instinct not merely for raw aggression and the mys-
terious will to do hurt that resides, for better or worse, in the
human soul, but for suggesting the incontestable justice of such
an instinct . . .
—Joyce Carol Oates, "On Mike Tyson"

WHEN HE BURST OUT of his corner and charged his opponent like a
bull chasing down a matador; when he stalked the ring in those plain
black shorts, short black socks, and black shoes; when he completely
overwhelmed and annihilated his higher-ranked, better-dressed opponents, Mike Tyson joined, for me, a long line of black heroes and icons;
so I suppose I've always been a bit biased toward his side of the story
of that one time in the ring, when he lost control, and bit off another
man's ear.

This kind of savagery I could forgive, even understand, if only
because I was pretty sure that it didn't mean what a lot of people
thought it meant. Or at least because I wanted to believe that the
answers to the question, "Why would he do that?" were perhaps more
complicated than they seemed.

////

JOHN HENRY, THE STEEL-DRIVING man, that mythic railroad man of American folk legend and song, was probably my first black hero. We owned a storybook version and, as a boy, I loved to revisit this tale of superhuman strength and endurance, loved to hear about his triumph over the train and technology, man over machine; and I still associate Henry and his ultimate sacrifice with the railroad and western expansion.

At my elementary school library I checked out all the biographies I could find because I liked "true" stories, and most often these were the stories of dead presidents or of black baseball players. I consumed them all, under the strict Dewey Decimal order of our librarian, Virgie Pine. But I also found, at school and our public library, biographies of Martin Luther King, Harriet Tubman, and later, Malcolm X; and despite my committed hatred for his team, the Lakers, I even read and appreciated Kareem Abdul-Jabbar's autobiography, *Giant Steps*.

It was through these books that I first I fell in love with heroes like Hank Aaron and Willie Mays, men who I'd never actually seen play baseball, the game that made them famous, men who'd overcome poverty, racism, and other degradations to succeed and thrive. It was probably through these books that I first learned about the racism and oppression in our country. I'm not sure how to explain this hero worship. I was a white middle-class kid growing up in a midsized Midwestern college town, a kid growing up with the remnants of generations of racism.

Granted, my hometown, Lawrence, Kansas, painted over any longstanding racial tensions with stories of a rich and proud abolitionist history, a town where people occasionally found hidden tunnels, caves, or passages beneath their home, places where runaway slaves had been hidden after fleeing Missouri; but Lawrence was also a largely segregated town that had learned to hide its racial divides behind civic pride and a collective memory of the time our town was a town burned to the ground by pro-slavery forces lead by the "terrorist," William Quantrill, whose "army" of Missouri farmers also slaughtered 150 Lawrence men and boys in the streets.

Perhaps I'd unconsciously internalized some of this history. Perhaps my love for black heroes was the only way I knew how to push against the white privilege with which I was raised.

Whatever the reason, I loved black heroes and, for a while at least, I specifically loved black boxers. This was before the sport seemed so corrupt and antiquated, before it migrated to Pay-Per-View specials and was eventually replaced in the collective psyche with Mixed Martial Arts fighting, what now seem like more "pure" forms of sanctioned violence.

I was a little too young to appreciate Muhammad Ali in his prime, but I followed some of the classic heavyweight battles of the late '70s and early '80s and remember trying and failing to summon some love and respect for the Great White Hope, Gerry Cooney. It seemed like what I was supposed to do, given the dearth of decent white boxers; but mostly Cooney seemed like a big clown, a pale white punching bag for the real fighters, and nobody I hung around with gave him much love.

My favorites, though, weren't the plodding heavyweight fighters like Larry Holmes. I wanted speed and aggression, and it was the epic middleweight contests between Sugar Ray Leonard and Roberto Duran, "Marvelous" Marvin Hagler and Tommy "The Hit Man" Hearns—but *especially* the fights between Leonard and Hagler that captivated me.

In my book you were either a Sugar Ray Leonard fan (meaning you were a flashy showoff, mugging for the cameras, and probably also a Lakers fan) or you were a Marvelous Marvin Hagler fan (meaning you were a blue-collar pugilist, and a Boston Celtics fan, who didn't give a shit about the cameras and just wanted to punch your opponent's nose through the back of his skull).

There was no in-between. Hagler or Leonard. Bird or Magic. Make your choice. These distinctions were important to me, the stuff on which friendships were forged and strained, and on which a great deal of time was spent talking shit in the halls of school.

When Mike Tyson came along he somehow embodied, for me, a little bit of both Leonard and Hagler. He was part showman and part pugilist. Short of stature and quick on his feet, with a neck that blurred into thick ropy muscle, Tyson seemed carved from marble; but he was still a small heavyweight who didn't have the reach that other boxers did. An old-school street brawler like Hagler, Tyson moved Leonard-fast in the ring but thrived in close quarters, raining a storm of fists into his opponent's torso until he could find an opening to unleash his uniquely murderous upper cut.

Tyson was capable of pure aggression and sublime violence, but also capable of grace and skill, with an innate understanding of time and opportunity. He was also an actor and a shape-shifter, incredibly adept at playing the role of the charismatic barbarian. It's no wonder Joyce Carol Oates found him so fascinating, so existentially compelling, and dedicated so many pages to him in her brilliant collection of essays, *On Boxing.* Tyson gave masterful performances as both monster and myth, while somehow remaining oddly innocent, charming, and captivating; and I, too, found the mix of these traits to be completely intoxicating.

My friends and I used to gather to watch some of Tyson's later fights, a room full of middle-class white guys all bouncing with energy and angst, working ourselves into a frothing stew of misplaced aggression. We gathered to witness the carnage and we paid for it. Pay-Per-View, pay-per-violence. And more than once, Tyson ended the fight so fast we barely had time to bond over the shared celebration of his violence. It was like a premature ejaculation of energy, all of us gushing and cursing and collapsing, spent, on the couch cushions.

ROUND 2

When I fight someone, I want to break his will. I want to take his
manhood. I want to rip out his heart and show it to him.
—Mike Tyson

ON JUNE 28, 1997, during the rematch fight between Evander
Holyfield and Mike Tyson—a fight billed as "The Sound and The
Fury"—things did not go well for The Fury.

Tyson had already been beaten badly by Holyfield in the previous fight, suffering a TKO, or "technical knockout," in the eleventh
round after a sustained pummeling. That match had shown Tyson to
be vulnerable, and he looked every part of the sports cliché "a shadow
of his former self." He made excuses afterward, claiming Holyfield
had used intentional head-butts to cut and daze him when the two
fighters entered into a clinch. Most people believed Tyson had lost his
edge, had grown fat on the largess of his life or been corrupted by the
influence of Don King—all of which was true.

Tyson's complaints, however, were not totally without merit.
Holyfield had long been known as a master of the subtle head-butt, a
tactic that, while common, is hard to spot and potentially devastating.
A head-butt in boxing is not the exaggerated forward strike you see in
professional wrestling or movies, but a more subtle tactic of close-in,
hand-to-hand combat, a swift strike with the crown of the head to the
thin-skinned brow, cheek, chin, or forehead of an opponent. Done

correctly, discreetly, it can quickly disable an opponent by knocking him stupid or by causing swelling to the eye or excessive bleeding, which blinds the fighter.

Intentional head-butting is against the rules of boxing, a violation similar to a punch below the belt. The head-butt is considered a "dirty" tactic—just the sort of trick you'd think a man like Mike Tyson, not a man like Evander Holyfield, would use to gain an advantage over his opponent. Perhaps because of this, all of Holyfield's head-butts were judged to be "accidental."

As the rematch fight entered the second round and Tyson's furious efforts to slow the fundamentally sound and patient and head-butting Holyfield seemed fruitless—several clean punches failed to deter Holyfield's steady advance—the two men entered into a clinch, and another head-butt from Holyfield opened a sizable gash above Tyson's right eye. With blood streaming down his face, Tyson complained bitterly to the referee and admitted later that he was dazed and scared, feeling vulnerable; but the referee, Mills Lane, again ruled the head-butt was unintentional.

Angered by the head-butt and Lane's refusal to intervene, Tyson came out for the third round and unleashed a barrage of punches at Holyfield, but his rally barely fazed the champ. The two men clinched up, and again Holyfield head-butted Tyson, who, at this point, became convinced Lane wouldn't protect him. He was desperate, angry, and determined to defend himself.

Tyson, despite all his fury and bluster, spoke with a lisp, and his characteristic high-pitched, nasally voice made him sound like a man-child, a curious mix of innocence and aggression. Tyson always, always fought as if he'd been beaten back into a corner and told to stay there. Like many people, I'd come to love Tyson for his naïve ferocity, for the brutality with which he dispensed opponents, often exploding as if he'd been unchained and turned loose from his corner. He didn't just beat his opponents; he went out there in his black trunks, black shoes, and short socks, and he humiliated his opponents. He destroyed them. We loved every terrifying second of the carnage, especially when he threw his upper cut, a punch that seemed capable of decapitating a man.

Most of us wanted Tyson to destroy Holyfield, to show the world that Iron Mike was still a force to be feared. But something was wrong from the beginning. Tyson wasn't himself, wasn't the Fury we expected. Instead, he became something else entirely, something much worse— a mirror, a reflection of our own bloodlust, a vessel for our collective savagery. Tyson lost the artistry that made his brutality beautiful. It disappeared into the fog of fear.

Lane separated the two men, and the fighters exchanged a few punches before locking up again. As they did, Tyson spit out his mouthpiece. When Holyfield's head came up, Tyson twisted his neck, tucking into the side of his opponent's face almost as if to kiss him on the cheek or nuzzle his neck.

Tyson opened his mouth wide. He bit down hard on Holyfield's upper right ear, severing the helix, the outer section of cartilage, the curve away from his head that defined his face. Holyfield jumped back, shoving Tyson, who spit the chunk of ear onto the canvas. As Lane tried to figure out what had happened, Holyfield hopped around the ring, gesturing at his head with his glove as blood poured from the open wound.

The boxers were sent to their corners. Everyone watched. Then, perhaps caught up in the moment and not fully aware of the extent of the damage to Holyfield's ear, or perhaps so sucked into the adrenaline of blood-sport that he was blinded to the reality of what was happening, Mills Lane allowed the fight to continue.

He wanted it to continue.

We all did.

After exchanging a few punches, the two fighters locked up again. This time, Tyson bit down on Holyfield's left ear—not as hard as the first bite but still hard enough to cut and leave a mark and send Holyfield jumping back, flailing his arms hysterically and pointing cartoonishly first at Tyson and then at his own head with his red gloves.

Lane again sent the fighters to their corners and finally ended the fight, waving his arms in the air and disqualifying Tyson, who exploded in rage and rushed at the Holyfield corner, throwing punches at anyone who got in his way. Soon, the ring was flooded with thick-necked sheriff's deputies in beige uniforms.

As Tyson exited the ring, boos erupted from the crowd. The whole place seemed to surge and pulse with adrenaline. Tyson bulled his way toward the exit, and a fan threw a water bottle at him. Tyson had just bit the ear off of another man and this fan threw a water bottle at him. What was he thinking? . . . Tyson jumped over the barrier, charging into the crowd, screaming profanities and pointing at people, raging at anyone near him. Members of his entourage and security personnel dragged Tyson out of the stands and pushed him toward the exit.

Unable to look away from the television screen, I watched the spectacle unfold from a safe distance with my friends, all of us gathered in the living room, drinking beer and eating chips; and when it all happened, time seemed to bend, as if we were all rendered in slow motion. Our mouths twisted into strange questions. *What? Did he? The ear?* And the images floated like party balloons. Red gloves. Black men. The bloodied helix. The white canvas. We curled into the moment, and it pulled us down. A dream? *This can't be happening, we thought. This can't be real.*

I watched the fight again and again on YouTube ten years later, after watching the Tyson documentary. I was facing my own battles, my own fears and still struggling to come to terms with my violent outburst a few months before at that party. I suppose I was trying to normalize the savagery in Tyson as a way to apologize for or confront my own.

In those days I spent the bulk of my time caring for my infant daughter, carrying her around on my hip or cradling her in a one-arm "football" hold as I made food or did the dishes. She also liked to ride around on my shoulders and would clutch tightly to my ears, using them as her handles to keep her balance. She loved my ears and regularly begged me to do my "ear trick," where I fold my oddly pliable outer ear down and stuff it into the hole, where it will sometimes stay for a few moments, until I flex a facial muscle and it pops out. She squealed with delight every single time.

Given a free moment or two, I'd sometime watch videos of the fight and think about that ear and what it was like to be close enough to touch Tyson's unchained violence. Most of all, I wanted to see the severed helix, that outer curve of the ear lying on the canvas or cupped in a white towel, maybe nested in a bucket of ice.

I'm not sure how to explain this fixation, this morbid fascination, except to say that it seemed connected to my thoughts on savagery and the line between human and animal. I tried to imagine what it would feel like to bite down into an ear, to feel the skin and then the cartilage give beneath your teeth, the crunch, the rush of blood, the metallic taste on your tongue.

ROUND 3

I just want to conquer people and their souls.
—*Mike Tyson*

DAVID LYNCH, THE MIND behind such cinematic creations as *Twin Peaks, Eraserhead, The Elephant Man, Dune*, and *Lost Highway*, also wrote and directed 1986's *Blue Velvet*, a movie that changed forever the way I thought about savagery and intimacy, and a movie that was released in the same year that Mike Tyson beat Trevor Berbick to become the youngest heavyweight champion in history.

The movie begins and, perhaps you'd rather linger in Van Gogh's yellow fields, but instead you unwittingly stumble into Lynch's uniquely twisted vision of the world. At first, you're simply watching a young man walk through a field of overgrown grass, and you have no idea what to expect. Perhaps you watch him closely, this angular, pasty-faced man in Lumberton. Everytown, America. Innocent but curious. A man, but just barely. Jeffrey Beaumont. He's cutting through this overgrown lot after visiting his dying father in the hospital. He's not expecting any complication to his life. He has no idea how quickly things can change, how one small discovery in a field of secrets can crack open his world.

Jeffrey finds an ear in the grass. A human ear. Severed from the skull, it becomes like a window, or a rabbit's hole. Lynch said of this scene, "I don't know why it had to be an ear. Except it needed to be an opening of a part of the body, a hole into something else. . . . The ear sits on the head and goes right into the mind, so it felt perfect."

THE BRILLIANCE OF THE scene is captured in the dilemma it hands off to the audience: The question put before them is "What would you do?" It implicates us in everything that follows—the twisted search for answers that leads to Isabella Rossellini and the infamous scissors, Frank and the mask, Jeffrey hiding in the closet, the car ride, the dancing, the drinking, the kidnapped boy, Dean Stockwell, nitrous oxide, and Roy Orbison—all of it spinning madly out of control. And it all begins with curiosity, attention to an ear and a lingering question: What would you do? Could you stop yourself from falling, too?

It's strange how a subject overtakes you. This thing that Jeffrey cannot leave alone, this thread he cannot stop tugging, makes everything come undone. Lynch's camera takes you down into the ear, and you fall into the darkness. You wonder, what would you let someone do to you in order to keep your child alive? With the scissors? Your life is never the same after this movie.

Again and again, I fall into the severed ear in this opening scene, this salvo across the bow of expectations, and I disappear into its rabbit's hole. This ear replaced Van Gogh for me—the easy, comforting allusion of aberrant emotion Lynch forever after corrupted and complicated in ways that are so much more unsettling than the anecdotal violence of Van Gogh's severed ear. Now, when I hear "severed ear," it's not golden sunflowers and unrequited love I think of, or even a drunken quarrel between rival artists, but, instead, the mess of Mike Tyson or David Lynch's particular brand of savagery.

ROUND 4

To come to a scene and you see a fellow human being ripped apart,
I feel for that.
—*Officer Frank Chiafari, Stamford Police Department*

TRAVIS, A 200-POUND, FIFTEEN-YEAR-OLD chimpanzee, lived in
a private home in Connecticut for most of his life. His owners, the
Herolds, operated a tow-truck business, and Travis used to ride along
in the truck to help stranded drivers. He was something of a local
celebrity, an animal that acted the role of family member and friend.
He drank wine, ate steak, and enjoyed many of the finer things of
human life; but it seems Travis was also a sad chimp—perhaps even a
depressed, anxious, and fed-up chimp, frustrated with his own brand
of captivity, tired of the expectations that he be so tame, so unnatural.

On the last day of his life, February 16, 2009, Travis was particularly agitated and had been given Xanax for his rage. It didn't help
much and the situation had escalated. He'd gotten outside the house,
out of control, and he was angry, storming around the yard. Maybe
Travis felt flooded, too, and couldn't find the numbers to count himself down.

His owner, Sandra Herold, had called her friend Charla Nash
as a last resort after she couldn't get him under control. Travis knew
Charla and seemed to trust her. She'd always had a calming effect on
Travis, and Sandra thought she might be able to help him relax.

Charla pulled up to the house and stepped out of her car. All she wanted was to get Travis back into the house, into his cage, to get him to calm down and feel safe again.

But Travis had crossed over. He'd broken character and become animal again. He'd gone savage. Charla barely made it a few steps before Travis attacked her in the driveway. By the time he'd finished, he'd bitten or ripped off Charla's nose and lips, her eyelids, part of her scalp and most of her fingers. One of the first responders on the scene said he couldn't believe an animal had done that and said it looked as if her hands "went through a meat grinder." For his part, Travis had suffered a stab wound to his back when Sandra Herold plunged a steak knife into him in a futile attempt to stop the attack.

Officer Frank Chiafari was among the first responders to the grisly scene. He'd known Travis and liked him, and probably thought of him as a mostly harmless but charming pet. But as Chiafari pulled into the driveway, though, Travis was no longer endearing, no longer harmless; he'd become something else entirely, a manifestation of savagery unchained, the worst side of nature. He had become chaos and disorder, an instrument of mayhem. He had crossed over to the savage side—or perhaps he had simply returned to the state from which he came, to the moral indifference of nature.

Charla lay semiconscious and horrifically maimed on the ground, and Chiafari knew he had to do something. But before he could even get out of his car, Travis approached, screeching and howling, and he knocked off the side mirror as if "it was butter," grabbed the handle and yanked open the door. In some awful approximation of a horror-movie scene, Travis stood there covered in Charla's blood, opened his mouth to shriek and bared his teeth.

Chiafari shot Travis four times with his service revolver. He stumbled away from the car, back into the house, crawled into his bed, and died where he had slept most nights.

Afterward, Chiafari was tormented by what he'd seen and by what he'd done. He couldn't visit a mall or an amusement park without being haunted by images of faceless women. He avoided zoos or the circus or any place where he might see a chimpanzee, an animal that Chiafari knew was capable of perhaps the most human and natural instinct of them all, extreme violence and savagery.

For some reason, Travis didn't touch Charla's ears.

Charla has since undergone a face transplant, but during her appearance on *The Oprah Winfrey Show* in 2009, Charla's ears were, in fact, the only feature that allowed you to recognize the thing on top of Charla's shoulders as human, the only recognizable facial feature on a head that looked more like an abstract sculpture of a head. Charla's eyes had been removed, and she drank through a hole in her face. She had only one thumb remaining.

Charla Nash could still hear just fine. And this, it would seem, was both a blessing and a curse. She still felt like the same person inside, and because she couldn't see or touch her face with anything besides her one remaining thumb, her main understanding of how she looked was gained by listening to other people's reactions. She could only hear what others saw—or didn't see—in her face, and she wore a veil to protect us. She said she still felt like the same person—still felt like a person.

"I just look different," Charla said.

ROUND 5

I love to hit people. I love to. Most celebrities are afraid someone's going to attack them. I want someone to attack me. No weapons. Just me and him. I like to beat men and beat them bad.
—Mike Tyson

At the end of David Lynch's cult classic film *Blue Velvet*, the camera emerges fly-like from protagonist Jeffrey Beaumont's ear as he's lounging happily on the lawn. The ear is a portal to another world, a world of the mind, an ecstatic reality that offered a frighteningly dark reflection of the reality in which we all were living in the early '80s. At this point in the film, we've found out, of course, that another ear, the severed ear Jeffrey found in the movie's opening scene, belonged to the dead husband of his neighbor, Dorothy, and that it was presented to her as a kind of warning, a demand that she give the amyl nitrate gas–huffing sadist Frank what he wants, whatever he wants, whenever he wants it. The severed ear, the one we descend into in the first scenes of the film, becomes a promise that Frank will do the same violence to her kidnapped son. But, in this final scene, this emergence, all of that is over now and Jeffrey's mind roams free, released from the pull of the ear and all it represents.

Lynch makes it all look so easy. But he never does give us the scene—the severing of the ear—just the husband's body in Dorothy's apartment and the lobotomized man in the yellow suit, his neck and chest painted with blood, as some kind of answer to our questions of why or how. We only see the aftereffects, the end of the savage play. In his first hit movie, *Reservoir Dogs,* Quentin Tarantino, however, shows us the act, or *most* of the act. He puts us in the room with violence and lets us listen to the soundtrack of savagery.

These images are catalogued forever into my consciousness and queue up whenever I find myself thinking about *Blue Velvet*, or Van Gogh, or even my favorite ear-biter, Mike Tyson. To this day, I am terrified of the actor Michael Madsen, who played the psychopathic Vic Vega, aka Mr. Blonde in *Reservoir Dogs*. If I saw him on the street, especially if he was wearing a black suit and black tie, I would not ask him for an autograph. I would not pose my children with him for a photograph. There would be no "Me and Mr. Blonde" selfie. I would run away. Because to me, he IS Mr. Blonde, inseparable from the role he played. I cannot see him any other way, not after that movie.

Like actor Kyle MacLachlan and his ear, who will always be *Blue Velvet*, Madsen is forever Mr. Blonde; and I can, with little effort, conjure up the scene from *Reservoir Dogs*: rookie cop Marvin Nash duct-taped to a chair, wounds blooming from his face already as Madsen, or Mr. Blonde, beats him. He sputters and spits, snot and blood running from his nose as he begs for his life, swearing he knows nothing about any set-up at the jewelry store. Marvin has children for god's sake! And Madsen . . . I mean, Vic Vega, or I mean, Mr. Blonde doesn't care whether he knows anything or not. He just wants to hurt him. He plans to torture Nash and he tells him this, tells him he doesn't even care what Nash knows and that he's just going to hurt him because he likes hurting people. And Marvin Nash is begging for his life, crying and spitting, almost pathetic, and you're thinking about his children at home, his wife waiting for him to return.

Mr. Blonde, with that cigarette dangling from his lips, pulls a straight razor from his boot—a fucking straight razor (what kind of psychopath carries a straight razor in his boot?)—and you just know something very very bad is going to happen, something grisly and memorable.

"Do you ever listen to K-Billy's Super Sounds of the '70s?" Blonde asks as he turns on the radio, twisting the knob to find his station. "It's my personal favorite."

And we hear the laconic deadpan voice of comedian Steven Wright as he introduces us to the song, and tells us that we are, indeed, listening to K-Billy's Super Sounds of the '70s.

This is when it comes: those first bouncy notes of that song by Stealers Wheel, "Stuck in the Middle with You," and the lyrics, "I don't know why I came here tonight/I got the feeling that something ain't right," and it's as if the music is saying what you're thinking. You don't know why you came here tonight. You've got the feeling that something ain't right, that, in fact, something is very wrong.

Blonde struts across the warehouse floor like a rooster, clutching the razor in his hand. He dances and shuffles, staring at his victim. Two stripes of red stream down from Marvin's nose, across the silver tape. He's breathing heavily through his nose and grunting beneath the gag. The music plays on. And Blonde dances toward Marvin, comically shuffling and wielding the blade like a paintbrush; he slashes him across the cheek. Then he grabs Marvin's face, holding the blade up, and studies his work. He considers his subject, his victim, his canvas for a second.

Blonde sits down on Marvin's lap, his back to the camera; sits there like he's going to kiss him or hold him, as if we are witnessing a tender moment between the two. But then he reaches over Nash's head with the razor. We see Blonde's back, the arching arm, the reach. But then the camera turns away, gazing up above a doorway where the words WATCH YOUR HEAD are spray-painted.

Blonde leans over and slices from the back to the front, severing Marvin's right ear. There is no reason for this. He just cuts for fun.

The camera pans back as Mr. Blonde stands up holding the severed ear and looking at it in his fingers. Nash screams in agony. Blonde waggles the ear around, talking into it, "Hello? Hello?" as Marvin makes more noise and twists uncomfortably in his chair. Blonde lets him squirm.

The song fills the warehouse space, rising up like a breath, and then fading in an exhalation, a brief respite, as Blonde makes his way outside to retrieve a can of gasoline from the trunk of his car. The music dies as the door closes, and, outside, you hear the sound of children playing, invisible children somewhere in the distance. You hope the children are far far away, because, back inside the warehouse, that song and that scene still await your witness. And you want to go back in that warehouse. And you *don't* want to go back in that warehouse. And both of these things are true.

The music swells up again as Blonde reenters the warehouse, carrying the gas can. And it's as if the song only exists in that room, as if it is a product of the situation, a spawn of the moment. To this day, I cannot hear that song without this scene replaying in my head, without seeing that warehouse and Marvin Nash taped to the chair. As with Madsen, I cannot separate the song from its role in the film. And Tarantino has admitted in interviews that the song came first, before everything else, and that he knew all along, "Stuck in the Middle with You" was the perfect musical accompaniment for torture.

Back to the scene, and it's clear to you that Mr. Blonde is enjoying this. He dances around, clowning for the camera, and douses Marvin Nash with gasoline. It's at this point that Tarantino finally shows us the wound. He gives us the image of the hole. The round black hole in the side of Marvin's head, bloody and wet and grotesque. It feels almost pornographic. An ear without an ear. A hole in his head.

Clowns to the left of me

Jokers to the right

Gasoline drips off of Nash's head, pouring over his open wounds. Watching the scene I can almost smell the gasoline burning *my* nose, stinging in the cuts. And the music keeps playing, never stopping, bouncing in the background.

Here I am

Stuck in the middle with you

Marvin Nash spits and sputters, trying to get the gasoline out of his mouth, and I wonder about it pooling up in his earhole, settling into the canals, draining into the cochlea and drum. I think about his ears burning.

Mr. Blonde is still dancing and spreading a trail of gasoline on the floor. He flicks the lighter open as Nash begs for his life, begs Blonde not to burn him; and just before you know he's going to burn the one-eared cop alive, just when you've completely forgotten about Mr. Orange dying in the background, swimming a pool of his own blood, the shots ring out. *Pow. Pow. Pow. Pow.* And Mr. Blonde drops dead, lumped on the floor, a casualty of savage justice.

Orange kills Blonde. And you've never felt so happy to see a man die. You wanted him to die. You wanted to stop him, but you were powerless to help. You wanted to hurt Blonde back, to torture him for the things he'd done, the things he planned to do to Marvin . . . or for the things you imagined he might do. And what is the line between imagining and desiring, between picturing and wanting? If you can conjure it, feel it jangling through your nerves, do you not on some level want it to happen? And isn't that the point?

We are capable of imagining and even of wanting to hurt. At least. If not capable of hurting, too. This is why the movie and, in particular, this scene succeeds—because it implicates you, the witness, in the savagery.

"*. . . he has the power to galvanize crowds as if awakening in them the instinct not merely for raw aggression and the mysterious will to do hurt that resides, for better or worse, in the human soul, but for suggesting the incontestable justice of such an instinct. . . .*"

This scene is the Mike Tyson of movie scenes. It makes us imagine Marvin's suffering and it makes us want to hurt Mr. Blonde. But even more troubling, the scene convinces us of the incontestable justice of this desire. He lets us feel the creeping horror, the suspense, and finally, the release, the ecstatic exhalation of breath when Mr. Blonde is shot dead by Mr. Orange. Tarantino gives us violence and vengeance that implicates the witness.

Of course our reprieve is a brief one, a momentary confrontation of our own instincts that passes quickly, like all adrenaline rushes. And after Blonde is dead, the music still plays on—*here I am, stuck in the middle with you*—looping endlessly through your thoughts like the jingle for a television commercial.

For the moment, Marvin the one-eared cop lives, if only for a little while longer, if only until Fast Eddie shoots him without a second thought, without even looking at him or his savaged earhole. Soon enough, plot takes over and moves the action forward, beyond this scene, we never see Marvin's ear again, never find it lying on a table, next to the radio, or beside Blonde's dead body, maybe clutched in his fist, or outside in the parking lot, covered in bits of gravel. The camera doesn't emerge Lynch-like from the ear at the end. We're just expected to forget about Marvin's ear. We're expected to remember the desire and the rush, the seductive release of justified violence. And I do. Those images linger. But I also always remember the ear, the sound, and the sense of doom. And that song—the bouncy, folk-pop imitation of Bob Dylan—and the way it amplified the savagery through its incongruity. *I got the feeling that something ain't right.* Because it's all wrong. The images don't fit the sound. They bounce off of it, standing out in stark relief. The song has now become infamous, inextricably linked to this scene, and many people can recall the same grisly images, the same Mr. Blonde dancing around with a straight razor, and feel the same pull of gravity, the same moral weight dragging them back to that warehouse.

ROUND 6

My style is impetuous. My defense is impregnable, and I'm just
ferocious. I want your heart. I want to eat his children. Praise be
to Allah!
—*Mike Tyson*

YEARS LATER, MIKE TYSON is only mildly repentant for biting off
Evander Holyfield's ear. Basically, he's sorry he was forced to do what
he did. A 2007 documentary on Tyson's life, narrated almost entirely
by Tyson himself, seems to support his claims that Holyfield used
head-butts for a tactical advantage in each of their fights and that, for
whatever reason, the referee in each fight mostly ignored the tactic.

Is it possible that Tyson became a casualty of his own image—the
mad-dog fighter, the Fury? If everyone else believed it, why *wouldn't*
the referees also buy into the hype of the Fury, the belief that Tyson
represented something animal and primal, while Holyfield symbol-
ized the Sound, the humble, soft-spoken gentleman boxer who would
never intentionally head-butt an opponent?

You can watch videos of the incident online. As bizarre and grotesque as it is to see the ragged tear on Holyfield's ear and the blood
pouring down his head, to watch Tyson spit out the chunk of helix,
and to think about what it would take to bite through flesh and cartilage, to sink your teeth into someone's ear, it is also strangely unsur-
prising. Normal. Even predictable and, in some ways, entirely justified

if you think about it. You might have done the same if you were in Tyson's shoes. Really, what would you do? How would you defend yourself from head-butts in the ring?

I often thought about how I could have handled things differently, how I could have avoided the confrontation with the guy at the party a few months before. I wanted to believe I could learn something from it, that I would somehow react differently, but I wasn't sure. I wanted to believe I could be a better model for my kids, that I could tell them that you don't solve problems with fighting, you just create more; but I'd also told them how, when a boy on the playground had called me "Scarface" because of the sizable mark on my cheek, I'd punched him in the stomach, "And he never called me that again," I said, wishing immediately that I could take it back.

I knew that was something they'd remember. Their father, as a child, punching another child. No matter what I tried, I seemed destined to broadcast mixed messages about what it means to be a good person. My daughter now asks me to tell her the story of the time I hit a boy on the playground or the time I slammed a guy against a wall. She wants the story. She wants to see me in the role I don't want to claim.

Aside from the impulse to self-defense, the instinct to bite is ugly and extreme, but it also makes sense. It's more natural than it might seem. As someone who obsessively chews his fingernails and his pens, who has watched his babies' faces contort with teeth-pain relieved only by chewing or biting down into something, who has seen a frustrated toddler bite because she can't do anything else, who has nibbled on his own daughter's ears, I can recognize that the urge to bite is not always an urge to hurt or rend flesh, but sometimes just an effort to find comfort and security, intimacy and escape from pain. It's often a desperate effort at self-soothing. My daughter used to bite me when we were playing happily, rolling around on the floor. She wasn't trying to hurt me. She'd just get overly excited. She was trying to hold me, to be close to me. She wanted a connection closer than touch, wanted to feel safe and secure. It was a love bite.

Tyson's no child, and I don't mean to infantilize him or apologize for him, or to suggest that his bite was a love bite; but perhaps it also wasn't motivated by hate, anger, and aggression. Perhaps it wasn't evidence of his savagery and insanity. Mike Tyson is not a great person. But he's also not a simple person. He's done a lot of bad things and hurt a lot of people, especially the woman he raped; but I believe his actions in that Holyfield fight were not so much those of a violent apex predator or a monster but instead the existential jaw-clenching of a lonely, frightened human being—desperate, violent, but not necessarily chaotic, aberrant, or inhuman. Is it possible to try and understand savagery without sympathizing or forgiving it?

I was never a big Holyfield fan, anyway. Fundamentally sound, he was also boring to watch and, as I've argued, kind of sneaky. He also didn't have the charisma that Tyson possessed. Besides, I'm not sure the blood or gore or pain of what Tyson did was any worse, any more savage than a Holyfield head-butt. Think about how much that would hurt. Tyson, too, was bloodied and probably concussed. Tyson, too, had cut flesh, bruising and swelling. But I understand. There is, in fact, something different about putting your mouth to another man's ear, sinking your teeth into another's flesh, something so intimate, the severing so desperate and personal, that it made us recoil and call Tyson an "animal," "psychotic," "savage," or, worse, "inhuman. " These were the easy answers. I hate easy answers.

Though I'm not sure we would admit it, I believe instead what burned viewers and fans (even the casual reader) most was the naked humanity, the unfettered vulnerability of what Tyson did, perhaps even the recognition of the incontestable justice of such savagery. What frightened us was his fear. What disappointed us was Tyson's weakness, even if we know that violence is often born of fear, insecurity, and weakness. We wanted him to be better than this, a vessel for our own vulnerability. We wanted him to save us from ourselves.

I suppose I want to normalize or humanize some of Tyson's savagery—the biting off of Holyfield's ear—as if I can scalpel it out of his larger body of violence like an unspoiled heart, but at the same time not apologize for the other kinds of savagery—particularly rape—of which he's been guilty. But I also think some of our reaction, our collective white middle-class revulsion at Tyson's actions, is a convenient intellectual lynching of a scapegoat.

Tyson himself once said, "When you see me smash somebody's skull, you enjoy it," and he was right. We did enjoy it.

But when he crossed that fuzzy line, we called him a savage, an animal. We vilified him as crazy and chaotic to make ourselves feel safer, to punish him for the very things that we loved about him and worshipped in the ring.

> *I'm on the Zoloft to keep me from killing y'all. . . . It has really messed me up, and I don't want to be taking it, but they are concerned about the fact that I am a violent person, almost an animal. And they only want me to be an animal in the ring.*

But we cut Tyson down for being an animal in the ring, and we forgave him, sympathized with him, paid him, put him in movies to play himself, objectified and named and worshipped him again. We made a cartoon of him. Tyson as phenomena—part real, part fiction—is as American and as immortal as anything. We love our violence, our monsters, and our stories of redemption.

> *I wish that you guys had children so I could kick them in the fucking head or stomp on their testicles so you could feel my pain because that's the pain I have waking up every day.*

As of this writing, Tyson is playing a narrativized version of himself live, on stage, for a touring Spike Lee production, a one-man show titled "Mike Tyson: Undisputed Truth," and he recently recorded a guest appearance on Madonna's new album. He's had other guest spots in several movies, including the box office hit *The Hangover*, which also featured his pet tiger; and if you do an image search for Tyson and "tiger" you can find images of Iron Mike posing in a lavish backyard pool area with his white tiger on a chain-leash. But it is Mike's choice to pose in nothing but a pair of saggy white briefs that strikes me as most odd—as if his tighty-whiteys aren't so tighty anymore. In other photos, Mike is wrestling the tiger in a wading pool and the underwear is all wet and droopy and kind of see-through. To me, the pictures capture the same sublime mix of strength and vulnerability that has always defined Tyson. Plus, they're just fucking weird.

In public, Mike still "goes off" sometimes and flips the switch, becoming the beast and cursing out a reporter or a heckler; and a recent newspaper headline in the *Daily Mirror* for a story on Tyson's "sex addiction," opens with the following lead, "Mike Tyson was the most feared boxer on the planet, a wild animal who literally tore chunks out of his opponents."

The audience for Tyson's new one-man show—held regularly at casinos—is often white and middle or upper class. They spend good money to watch their boy, Tyson, perform; and I'm sure there is some part of that audience waiting, hoping, wondering if he'll so something "crazy" or go "animal" again. Iron Mike is as much a construct of white middle-class angst and privilege—the objectified demon in the ring that we try to control, the contemporary King Kong on display, the humanized Travis—as he is a living, breathing, faulty human being capable of the most personal kind of violence and, even of naked, saggy-drawers vulnerability.

Several years after the Holyfield incident, in an interview after a warm-up victory against Lou Savarese and before a 2001 bout with Lennox Lewis, Tyson seemed juiced with rage, charged full of adrenaline as he barked at reporter Jim Grey a series of mumbled prayers to Muhammad. Grey, already looking ahead to Tyson's bout with Lewis, asked if the fight, which ended in a TKO after only thirty-eight seconds, had been Tyson's shortest fight ever.

Tyson, ignoring Grey's question, launched into a bizarre and impassioned monologue wherein he talked at least twice about having to bury his best friend, about how this fight was for the dead friend. He was grieving publicly, painfully, and you could see him focus his grief and his rage on Grey's question and on his role as Tyson the Savage, *the animal who literally tore chunks out of his opponents.*

You could almost see the switch flip, and watch his manufactured persona rising to the surface, as Tyson seemed to inflate with rage and mania, comparing himself to Alexander the Great, Sonny Liston, and Jack Dempsey, calling his style "impetuous" before eventually threatening to eat Lennox Lewis's children. He sounded crazy, like a man who'd lost his way and crossed over to the savage side.

I was struck, watching the scene, how similar it was to scenes from Werner Herzog's documentary *Grizzly Man*, where Timothy

Treadwell also seems to flip a switch and "turn on" his character, referring to himself as a "samurai" and a "kind warrior." Both men deliver these lines that are filled with angst and wonder, or macho violence, thoughts that seem to be bubbling up from a common core, from some molten mix of man and animal, an odd mix of humility and megalomania, rage and love. I also thought about some of the things I've said on the basketball court. Awful crazy things. Sick and disturbing things that frightened people. *Rip your head off. Shit down your neck.*

Lennox Lewis did not have any children when Tyson threatened to eat them, but the line, delivered at the end of an outpouring of grief over the loss of his friend, has become, in popular memory, another tag, another mark—further evidence that there is something deeply wrong, something dangerous about Mike Tyson.

Mike Tyson wants to eat your children.

Mike Tyson wants to kick your children in the fucking head.

Mike Tyson wants to stomp on their testicles.

Mike Tyson wants you to feel his pain.

ROUND 7

Tyson suggests a savagery only symbolically contained within the brightly illuminated elevated ring.
—Joyce Carol Oates, "On Mike Tyson"

NEARLY ALL OF US know the story of the painter Vincent Van Gogh, the tortured artist who cut off his own ear and mailed it to a lover. We keep Van Gogh's ear-story close as a kind of parable, a lesson or warning, perhaps a story of mythic and aberrant love. But if you're like me, raised in the cultural crucible of the 1970s and '80s, there is as much or more gravity in David Lynch's ear cradled in a bed of grass, or Mr. Blonde removing Marvin Nash's ear, or Mike Tyson biting off Evander Holyfield's ear, as there is in this old story of Van Gogh's desperate act. The story has changed, lost some of its force, but the archetype remains. Pop culture possesses, at its best, the power to replace Van Gogh with Lynch and Tarantino or Tyson, to adapt an old story of severed ears, savagery, and sacrifice to a modern context.

In everyday life, we don't think much about our ears or pay much attention to the ears of others; but as this scene from *Reservoir Dogs* reveals, ears have the potential as objects, as bodies themselves and parts of bodies, to shake up our measure of the balance between human and animal, between good and evil, and to challenge our understanding of savagery and, perhaps, of ourselves. We forget the ear until we can't shake its absence. We forget Van Gogh and remember Tarantino. We forget that our ears were one of our first sense organs to develop in utero, that sound was our first experience with the outside world.

We want to believe that we're not like Mr. Blonde or Orange or Tyson and that we're far from Travis and his brand of savagery, too, but try as we might, we cannot always remove ourselves completely from the animal urge to bite, to sink our teeth into something substantial, something firm but forgiving, something made of flesh and bone— especially when we are at our weakest and most vulnerable, or when we get punched in the face, our plans collapse and we just want to hold someone closer than seems physically possible, to consume that person and keep him inside us forever.

WE FORGET THE EAR until we see it severed again and again, until it becomes a pop-culture trope, both real and metaphor, both symbol and something tangible and terrifying. We forget the ear but not the story of its removal, not those lingering images of savagery and separation that leave us, *feeling that something ain't right*. And I'm willing to admit that I can't ever forget the ear, can't shake it as a symbol, because I've seen a necklace of ears, like a string of dried apples, kept as a trophy in an underwear drawer, nestled in a bed of white.

The father of a childhood friend stashed them there. We weren't supposed to be looking.

"You wanna touch 'em," my friend asked one day when we were alone in the house, standing in front of his father's dresser, the top drawer pulled halfway open.

"No," I said, and told him to put them back.

He told me that his father had cut the ears off of Vietnamese soldiers, men he'd killed in combat. He kept the necklace as a reminder.

I didn't ask of what his father wanted to be reminded, but I had a pretty good idea. I saw the man, the father with all his secrets, standing there, at the end of a long day. He tried to keep the ghosts at bay, keep them from rising. Tried to reconcile the necklace and the numbers. He opened the drawer and took out the ears, rubbing his fingers over them, worrying them down until they softened and bent to his touch. He looked at himself in the mirror above the dresser and tried to recognize the face he saw there.

As a child growing up in the '70s, to me the Vietnam War was mostly a knot of secrets that the fathers of other boys brought home with them. These fathers didn't talk much about how it was twisting them up. Perhaps they didn't have words for it. Most of my understanding of the war came through books, television, and movies; but when I saw that necklace of ears in my friend's house, I faced, perhaps for the first time, a harsh reality of war and savagery and torture.

I recoiled from the drawer, drawn down the hall, away from the ears. I didn't want to see anymore, didn't want to believe.

My father had avoided the draft, his number never called. But that day at my friend's house, standing in front of that dresser drawer, I understood suddenly that I was part of a generation of boys raised by men who'd done terrible things, men who killed and mutilated other people—often innocent people—for reasons they couldn't or wouldn't articulate; men who would become fathers and friends, football coaches and attorneys, bricklayers and ditch-diggers, Boy Scout leaders, elementary schoolteachers, professors, and writers.

By the time the song, "Stuck in the Middle with You," by Stealers Wheel was topping out at #6 on the Billboard charts in 1973, the Vietnam War was already winding down. On January 27 of that year, the Paris Peace Accords were signed and thus began the United States' retreat from the most costly—financially and morally—war since World War II. Nearly six months later, on July 9, my younger brother was born; and to me his birth signaled the end of my time in the spotlight and the beginning of battles with his existence.

My friend told me once that his father had been nicknamed "The Preacher" by his platoon in Vietnam. In the times I spent at their house on the golf course, I'd barely heard the man speak a word, so I never knew if the nickname was ironic or honest; but I was assured that he'd been an outspoken leader, often dispensing advice to the younger men and boys, and that he'd carried the biggest, heaviest gun, the M-60, all by himself.

I was already afraid of him. But something shifted that day. A new kind of fear arose in me, a fear for the future, a fear for all of us. I don't know if I would call it a gift or a burden. But that day at my friend's house I was given a vision, a truth that I would carry with me forever, a ghost that would come back to haunt me, rising up like a fever some twenty years later as I watched Mr. Blonde carving up Marvin Nash on the movie screen. For me, there was no Van Gogh anymore. There was no quaint story of aberrant love, no famous tale of self-mutilation. That was all gone. Now there was only Tarantino and Mr. Blonde and a necklace of human ears. There was only the recognition that man is more savage than any animal. And in the background, that song. Bouncing and jumping, *stuck in the middle with you.*

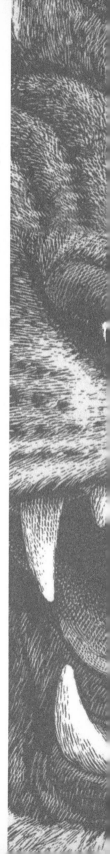

PART SIX

FATHER
AND
DAUGHTER

TRIPS TO THE ZOO

With our new habitats, we are trying to conceal
from ourselves the zoo as living evidence of our
natural antagonism toward nature; the zoo as
manifestation of the fact that our slow, fitful progress
toward understanding the animals has always been
coterminous with conquering and containing them.

—CHARLES SIEBERT,
"WHERE HAVE ALL THE ANIMALS GONE?"

OUR ZOO

When we first moved to Fresno, the Chaffee Zoo—located in Roeding Park, a mile or two from our house—was home to two chimpanzees that lived in a small habitat near the snack bar. Their enclosure was decorated with a sign reassuring visitors that the items of garbage strewn about the area—cereal boxes, plastic jugs, and paper bags—were in fact developmentally appropriate toys. This made me feel better since this was often how my kids' bedrooms looked; but the chimps didn't seem terribly excited about the toys or learning much from playing with them. They didn't seem terribly excited about anything. Actually, they seemed kind of pissed-off and antisocial.

Usually once or twice a week, I loaded my son (and then later, my daughter) in the car or the bike trailer and pedaled us to the zoo. After a short walk we'd inevitably find ourselves standing in front of the chimpanzees' habitat, a space that looked like one of those dirt-yards you see in Fresno with a pit bull chained to a tree and an overturned water dish, maybe a few patches of yellowed grass and a child's tricycle partially submerged in a mud puddle.

We spent a lot of time at the zoo and rarely saw the chimps. They stayed hidden from view, back behind the walls, through a dark hole. If one of them was out, he usually sat on a raised and shaded platform with his back to the viewing area. Occasionally one of them would stare out at us, looking passively curious.

I knew that an adult chimp could probably rip my arm off and beat me with it at the same time he was eating my face; but these chimps looked overweight and, frankly, kind of lazy. It was hard to imagine either of them capable of the sort of violence that a psychotic Travis had unleashed on Charla Nash. But, as it turned out, these Fresno chimps looked the way they did because they were profoundly unhappy.

Not long after we began frequenting the zoo, the chimps disappeared entirely and an article ran in the local paper with the headline "Grumpy Old Chimps," explaining that the chimpanzees had been at the zoo for a long time and had become increasingly disgruntled and quite depressed. They'd been acting out, it seems, and causing problems.

Instead of attacking a visitor or each other, the chimps responded to their sadness and existential angst and grumpiness with one of the few weapons at their disposal—excrement. They threw their shit at people.

It wasn't pretty. Visitors complained. Signs were posted. Calls were made. And eventually the chimps were shipped off to some kind of primate sanctuary retirement home (where poo-throwing is either accepted or otherwise mediated) and replaced the apes with a couple of giant anteaters that looked like dust mops with legs, and two giant rodents called capybaras, who had long orange teeth and a kind of sinister way about them.

Perhaps the chimps' sadness was too much for visitors to witness, too uncomfortable to acknowledge because they are so much like us; and I think it's true of most people that, if they were in a similar situation, confined to a zoo cage until they died or got shipped off to a retirement home, they, too, might express themselves similarly. I could almost see myself there. And if some dumb dad got too close to my cage, if he crossed the line to impress his daughter or something, I'd probably take a swipe at him. I might even try to pull him in with me. I might even throw some crap at his kids. And though the Fresno zoo was often a place fraught with contradictions—where our *slow, fitful progress toward understanding animals* often seemed to be a plodding march of martial law, defined by cages, control, and never-ending cycles of death and replacement—it was also, paradoxically, a place of great comfort and escape for my children and me. The zoo was a safe space, defined by the possibility of wonder, and a place where we could face our own fears and obsessions, or simply a place where we could get a "squirrel" ice cream cone and watch the sea lions swim anxious circles.

CONRADO'S TRIP

On the night of September 27, 1982, security guards from the Central Park Zoo twice evicted Conrado Mones from zoo grounds. He'd been lingering around for a couple of days, acting strangely and drawing attention to himself by crossing marked boundaries and fences, getting too close to the animal enclosures. "You have to get close to the animals," he'd told one zoo official.

When he was found on zoo grounds at 11:30 PM, well after closing time, he was quickly escorted outside and told not to return. But at 3:00 AM the next morning Mones was back again, loitering just outside the African lion habitat.

As he was led away from the cages he reportedly begged of a guard, "Help me."

THE CENTRAL PARK ZOO, like the Prospect Park and Bronx Zoo, are all owned and operated by the Wildlife Conservation Society of New York City, a massive organization responsible for a huge number of animals and visitors, not to mention some of the most prime real estate in the five boroughs.

All the zoos shared a similar original design, featuring a main square with a central exhibit—often the polar bears in the past or sea lions today—surrounded by different brick "houses" or buildings that contained cages and vitrines of animals. The "Monkey House," as you might imagine, housed the zoo's collection of various monkeys and primates. Today these old "houses" have either been converted to diorama-focused exhibit spaces with smaller animals, insects, and reptiles, or they've been converted into zoo administrative offices, gift shops, and cafeterias; and the animals have been moved into larger enclosures spread around the zoo grounds. These changes began in earnest in the mid to late '80s, perhaps in response to some troubling after-hours incidents at some city zoos, but also in response to new understandings about the physical and psychological effects on animals of cramped confinement in "unnatural" habitats.

SIEBERT'S ZOO

In his somewhat controversial essay, "Where Have All the Animals Gone? The Lamentable Extinction of Zoos," (which could probably also be called "In Praise of Cages") from the May 1991 issue of *Harper's*, Charles Siebert argues that zoos—especially the old-style zoos with their metal cages and cramped confined habitats—serve an important function, one that has been mostly lost with the recent trend toward these more "natural" looking larger habitats and animal "friendly" zoo designs, an aesthetic that tries to mask the ethical dilemmas inherent in any zoo experience.

Siebert says, "These are places we've designed to make *ourselves* happier about our continued keeping of (animals)," and these new zoos are designed to also help us feel better about the human destruction of wilderness and natural habitat around the world by recreating it in a suburban park:

> *They've (animals) passed so quickly from being curiosities to being*
> *scattered sympathies: from being the captive representatives in*
> *crude cages of an extant, flourishing wilderness to being the liv-*
> *ing memories of themselves in our artful re-creations of a vastly*
> *diminished one.*

What looks like an African savannah or a realistic bear den is really just a slightly larger cage, one that in reality is nothing like the animals' true natural habitat. The lions and tigers and bears aren't any happier, necessarily. They're still caged in a zoo. The new cages just make *us*, the visitors, feel better about viewing the animals in captivity. They're designed to be more aesthetically appealing and ethically inviting to human visitors, thus allowing us to forget that many of the habitats artfully recreated at the zoo are, in reality, being destroyed by human industry.

The new-style zoos remove the bars entirely in many cases, replacing them with moats or electrified wire, even just a large pit. Often you are above the animals, looking down on them. Often you can just step or climb right over a boundary and get up close to an animal for that one unforgettable photograph. But more often, you are far away and can't even see the animals hidden away in their habitat and you have little to no intimate interaction with them, and this for Charles Siebert is the fundamental problem with housing animals in "habitats" versus "crude cages." We've lost something fundamentally important about a trip to the zoo:

> *With all of them standing far from us, in their new habitats,*
> *we are no longer confronted with ourselves. We no longer have*
> *to look up close at who we are . . . and if we can't look at them,*
> *we can still see them in what's left of the wild—a habitat, a*
> *theme park we view like dismissive gods with a passing wave*
> *from the monorail.*

And sometimes, instead, perhaps we have to leap from that monorail in order to truly confront ourselves.

BRONX BEARS

In September of 2014, when I visited the Bronx Zoo, it was a long dark train ride beneath Brooklyn and Manhattan from Flatbush, where I was staying with a friend; but as the train emerged from the tunnel up onto the elevated rail, the sun streamed in the windows and soon we arrived at a stop just a short walk from the Wild Asia entrance gate. I'd come to try and find some answers, or perhaps some new questions.

I'd tried, somewhat halfheartedly, to find David Villalobos, but as a condition of his plea deal for the trespassing charge, he was confined to a mental health institution; and of course such places, because of patients' rights, won't disclose if a patient is staying there or not. Part of me also knew that, by now, his understanding of his leap, and the larger meaning of it, would've probably changed, would've been colored by consequences and judgments, recriminations and regret. It would have been the subject of therapy sessions. Part of me wanted David to stay in the realm of myth and mystery.

Part of me *needed* him to stay there.

My biggest fear: that I'd find him, talk to him, and he'd tell me his leap didn't mean anything, that it was all a mistake.

Of course it was a mistake, but it is often in such imperfect expressions of oneself that we often find the truest vulnerability and something close to art, or at least ineffability.

FOUNDED IN 1899 AND expanded (or revised) many times over the years, the Bronx Zoo is a massive sprawling heavily wooded park with a central square, featuring the similar kind of classic 1930s era Beaux-Arts architecture of the Prospect Park and Central Park zoos; but the Monkey House has been emptied of monkeys and the Horns and Hoofs building now houses the Zoo Education Center. As with the Prospect Park Zoo, a large sea lion exhibit occupies the central spot in the main square, where the animals swim frantic loops, breaching and diving obsessively. Walking past them and a rainforest greenhouse building, between two mountainous bronze rhinos, their rumps, tails, and noses polished golden by the hands of children, you next enter the main rotunda building, a stunning brick dome with moons of opaque glass circling the ceiling.

Here one of the zoo's rhinos has a featured spot, across from the Komodo dragon. But most of the animals have long ago been spread out into "regional" or "thematic" mini-parks, Africa, Asia, etc., some of which, like the Congo Gorilla Forest or the Wild Asia exhibits, require separate entrance fees. I paid the $34 for the Total Experience package, allowing me access to everything at the zoo, but that didn't get me past the low split-rail fences, the moats and hidden wires that separated me from the animals. On the day I visited, I was joined by mostly small children in strollers being pushed by their nannies or grannies. The older children were all in school.

Perhaps because the weather was temperate, sunny but not humid, a slight breeze blowing through the park, the animals were out and active, visible and photogenic. I didn't have to search too hard to find two grizzly bears playfully wrestling in their pool, and I couldn't help but think how much healthier they looked than Fresno's grizzly bear, Betsy, had looked near the end of her life. This bear habitat was at least ten times the size of Fresno's and included a couple of pools, a waterfall and stream, and a huge granite rock. Part of me thought that Siebert was just wrong, that the animals were actually happier in these new-style habitats, and thus the whole experience was less existentially troubling. According to the sign, they had at least five bears and, besides the two wrestling in the pool, I could just barely see another one on the hillside above, mostly hidden from view, far away from the visitor viewing areas. It felt, honestly, like I'd taken a step into another world, that I'd been given a brief window in the natural lives of these bears—even if it was all artifice and clever subterfuge.

My positive feelings faded pretty quickly when I walked around the corner and down the hill, where just on the other side of the grizzly habitat, the polar bear, a male named Tundra, stood on the banks of his pond, wagging his head back and forth, pacing nervously.

His nose looked a little gray with age and the bear seemed hot and tired. His paws were enormous, the size of dinner plates, but he looked slow and old, not even capable of an attack—even if I knew this was a lie. He shuffled his paws, wagged his head, and repeated. Shuffle. Wag. Repeat. Nearby, a plastic ball and ring floated in the pool, making it look a lot like any backyard pool in California where the kids have left the water toys out.

A deep concrete moat separated visitors from the bear, so it would be difficult (though not impossible it would seem) to jump into the pool. Water trickled out of the pool, down one side, and disappeared into the moat. The bear shuffled on the bank, looking like he was afraid to dive into the deep end.

Tundra did not have the presence of a predator so much as he did of a nervous old man trying to cross a busy street; and all the people passed by, pausing briefly to consider his predicament, before quickly moving on to something more entertaining.

BINKY BEAR

Binky, a polar bear living in the Anchorage Zoo, became something of a cult hero when he mauled an Australian tourist. The woman, posing for a photograph with Binky, crossed a fence and got up close to his cage. Before she could finish her pose, Binky grabbed her leg, pulled it through the bars and bit into it, breaking her leg and ripping off her shoe in the process. No punitive action was taken against Binky, however, as it was the woman who'd disobeyed the rules, ignored the signs, and crossed the boundaries. Not only was he not punished, but Binky also became something of a cult figure in Anchorage.

A famous picture of the bear carrying the shoe around in his mouth was printed on bumper stickers and T-shirts with the words SEND ANOTHER TOURIST: THIS ONE GOT AWAY. Binky didn't give up the shoe for four days. The woman, I'm sure, didn't give up her memory of the whole incident nearly so quickly. Broken bones take a long time to heal and the psychic scars must have run deeper and longer. But the truth is she was lucky to lose just her shoe.

A polar bear can smell a seal through five feet of solid ice, can track the scent of meat or blood for miles, and are one of the few bears that have been known to repeatedly prey on humans. Lately they've become the face of global warming, the face of Coca-Cola and of the Nissan Leaf electric car, mainly because they're so damn cute and so visibly endangered. They're like the supermodels of global warming.

In the Nissan commercial a polar bear, pushed off his ice floe by gas-guzzling SUVs, gently hugs a driver of the electric car, as if to thank him for his eco-conscious purchase. But this sweet and cuddly commercial persona ignores the fact that intimate polar bear–human interactions, especially polar bear–human interactions in zoos, rarely go well for the humans. In reality that polar bear, chased down from his natural habitat by starvation and global warming, would have most likely severely mauled, killed, and eaten the man right there in his driveway next to his eco-friendly Nissan Leaf. That bear would have made a buffet of his rib cage. It's not a pretty picture, but polar bears are rarely the cuddly mascots we make them out to be.

Your chances of surviving a polar bear–human encounter would be even worse should you be unfortunate enough to find yourself inside a polar bear's cage, rather than in your driveway. The description of a 1972 incident at the Toledo Zoo is indicative of the general tenor of such encounters:

> *A 19-year-old boy who either jumped or fell 15 feet into a polar bear pit at the local zoo last week was alive when he landed and was immediately attacked, killed, and eaten by four 800-pound bears.*

Though there was no real evidence to support the claim, speculation was that the boy, Richard Hale, was on "drugs" at the time of the incident. His death was thus ruled "accidental" and not an intentional suicide by bear.

Often those people who put themselves into intimate proximity with polar bears or other predators are described as deranged or insane or on drugs; and at least part of the reason for this is because it is difficult for us to imagine why anyone possessing all his rational faculties would jump into a cage with four 800-pound apex predators. But on some level we also understand, or we can at least recognize the appeal of the animal itself—perhaps even more than a tiger. We know the polar bear as mascot, polar bear as icon, cartoon, Coca-Cola–loving, white and cuddly. But I can't help but think of the polar bear as the page—blank, white, and infinite. Perhaps the polar bear is also possibility, grace, and the *ungraspable phantom of life*. Maybe on some level, each of us can recognize the urge to leap toward this, or toward the idea that a polar bear represents, this drive to confront annihilation in the cage. It's hard to intellectualize it, but we can feel the pull. Can't we? I want to believe I'm not alone in my Ahab-like identification with this urge.

CONRADO'S TRIP

After Conrado Mones was evicted that last time by zoo security he must have waited nearby, biding his time, perhaps hoping someone or something else might stop him, might save him. But nothing stopped him from breaking into the zoo again in the predawn hours. Mones scaled fences, crossed all posted boundaries, evaded security, and found his way to the polar bear habitat.

Charles Siebert says that "The old city zoo was designed, as a visit to an art museum is, to invite our immersion in the works and have us edified by them in some way."

The zoo grounds would have been quiet, dark, and peaceful at that hour, a respite from the noise of New York City, a small oasis. No screaming children. Lions sleeping, birds roosted for the night. I can almost picture the moon as Mones climbed three more fences and lowered himself into the polar bear enclosure. I can see the bluish light reflect off the water, the white curves of a sleeping bear—like an impressionist painting of nighttime at the zoo.

Perhaps Mones had a few moments inside, a few seconds, even minutes to feel the tingle, the surge of adrenaline that one gets in close proximity to an apex predator. Or maybe he was attacked instantly, as soon as his feet touched the water. The bear may have been waiting for him, or just waking from sleep, confused by the intruder. It's hard to say for sure as there were no witnesses to Conrado's leap.

We know, however, that Mones was eventually killed and partially eaten by a 1200-pound bear named Scandy. They found his body later that morning. According to zoo officials the bear was playing with the body, tossing it in and out of the water. Scandy, a bear described by officials as "friendly and gentle," was just doing what was natural and normal for a polar bear and, thus, no retaliatory or punitive action was taken.

SIEBERT'S ZOO

Zoos, according to Charles Siebert, serve both an educational and an existential function. They exist at least in part to remind us of the power dynamic between humans and animals as well as the ethical obligation we have to them, the responsibility we carry for their habitat destruction and even their species extinction. Zoos are supposed to be uncomfortable. They're supposed to make you feel uneasy about the line between animals and us. They're supposed to confront, challenge you, and even threaten to destroy your sense of security and superiority. Perhaps they're even supposed to be a little dangerous, especially if you get too close to the cage. Granted, Siebert's view of zoos and of animals is perhaps a bit darker than some. He definitely has his own ideas about why people visit zoos, or why they should visit zoos:

> *People visit zoos, I think, to have some telling turn with the wild's otherworldliness; to look, on the most basic level, at ways we didn't end up being—at all the shapes that a nonreflective will can take . . . visiting a zoo and staring at animals can somehow stay us a while, reinvolve us in the matter of existence.*

I sometimes wonder what it would be like to visit a zoo with Werner Herzog and Charles Siebert—just the three of us staring at bears staring at ourselves, considering together the sublime beauty of a "nonreflective will." Perhaps we'd all have ice cream cones. Perhaps we'd all be licking and pointing. Perhaps we'd be laughing at the chaos and murder at the heart of every interaction we have.

For Siebert and for Herzog, perhaps even for Schopenhauer, to contemplate a great predator like a bear or a tiger is at least in part a fairly selfish exercise in self-examination, a kind of confrontation of one's innermost being, or as David Villalobos said, "a test of one's natural fear." It is not a gesture of sympathy or altruism, or a selfless sacrifice, but a deliberate effort at experiencing otherworldliness.

I love our local zoo in Fresno. But it is not always an easy place and very rarely feels otherworldly or a place where we can *reinvolve ourselves in our own existence*. Tucked into a corner of Roeding Park, near the Tower District and Downtown, the Chaffee Zoo is a mix of old and new styles of zoo, an oasis from the summer heat, and a place where I've spent countless hours wandering with my children. But it is an urban zoo, a mix of the old-style cages and the newer "habitat" focus, and its small size means the animals are crammed in next to each other.

Mostly I visited the Fresno zoo with my daughter since, outside of some part-time preschool, I've been her primary daytime caregiver for most of her childhood. My schedule as a professor has allowed me to supplement what we couldn't afford to cover in daycare; but more importantly it allowed me time to be there for my kids, time to take them to the zoo and do other things. The Chaffee Zoo, however, is located in a park you don't want to visit after dark, a park where you can buy sex in the bathroom or trade it for drugs, a public park you even have to pay to visit during the day.

We still bought a membership every year and made regular trips, often on Fridays, our "daddy-daughter days," and for her, part of the deal was always the "squirrel" ice cream cone from the snack bar.

She loved to wander the zoo grounds, chattering happily, licking her chocolate-vanilla cone down to the melting stump as we visited all of her favorite "aminals." I did not grow up in a city with a zoo, but I spent many hours at the Natural History Museum at the University of Kansas, wandering the dark hallways lined with vitrines and dioramas, staring at the snakes and skeletons, the Tar Pit tableau, and the taxidermied polar bears.

In a city with no real museums and few parks or other opportunities for entertaining children, the Fresno zoo was a great investment. My daughter and I loved our time there together. When she got older she spent weeks there in the summer for zoo camps and, I'm sure thanks in part to her experience at the Chaffee Zoo, developed a deep and abiding love for animals, dreaming at one point of being an "animal rescuer survivalist."

The zoo was a safe space for my daughter, a place where I could let her run free and didn't have to worry about traffic or human predators. When she started walking more, I'd let her be my tour guide, taking us around the zoo grounds in whatever chaotic or frantic order she chose. The nice thing about being a zoo member is that we never felt pressure to see everything during a visit. We could always come back the next day. And when she got tired of the elephants or the Rainforest or of trying to find the black-footed cat, the Reptile House, and feeding the giraffes with their creepy prehensile tongues, she'd inevitably demand, "Up. Up. Howa." (her word for "hold") and I'd haul her up to my hip or my shoulders, where she loved to ride, her hands gripped tightly to my ears like they were safety handles.

THE BOY'S TRIP

On November 4, 2012, just a couple of months after David's leap, a woman and her two-year-old son visited the Highland Park Zoo in Pittsburgh, Pennsylvania and decided to stop by the African painted dogs exhibit. At the observation deck overlooking the dogs' paddock, the two of them would've peered over a railing into an enclosure that housed eleven painted dogs.

Fall had come to the zoo, and perhaps there was a red, yellow, and brown blanket of leaves on the ground of the large enclosure. The painted dogs, averaging between thirty-five and eighty pounds, might've gazed up at the mother and son with their exaggerated eyes, big rounded upright ears, and markings similar to those of a calico cat. Wide-eyed, they would have stared at the boy staring down at them. These dogs, in pictures or even from a safe distance, are undeniably cute, not obviously threatening or intimidating. They have ears like the oversized, pointed ears on one of our family dogs. I can imagine my daughter saying, "They look like Neko."

The boy—the mother's only son—was a happy, curious child, certainly the center of the family's life, and like all little boys at the zoo he must have wanted a better view of the animals. He wanted up, wanted to gaze over the edge, into the cage. He wanted to be closer. Closer to the funny-looking dogs. My own daughter is currently obsessed with dogs and it's easy for me to imagine her wanting to get closer to them, too. I can see her straining to see over the edge, into the cage.

The mother lifted her son up, setting his feet down on a wooden railing high above the dogs. I've done the same with my own children a hundred times at various zoos or other attractions. The boy would have had a commanding view of the enclosure fourteen feet below. I can see the mother's hands gripped tight to his waist, trying to keep him still up there, holding him back from the brink.

Immediately below the boy's perch, the zoo had installed a small net to catch cell-phones and cameras dropped by visitors, which was apparently a common occurrence. When the boy lurched forward, squirming out of his mother's grasp, he fell and bounced off this net before landing in the dogs' pen.

Descriptions of the boy's death are difficult to read. It's hard to even name him here on the page, hard to talk in specifics. The accounts suggest he was immediately set upon by all eleven dogs and, according to some reports, the animals quickly "tore him apart." One dog refused to leave the boy's body and was shot and killed. The other dogs were all "quarantined," and eventually replaced at the zoo with a cheetah, one of the fastest predators on the planet. The coroner's report says the boy was alive when he landed and that he died from blood loss. But perhaps the most disturbing accounts are those of his mother's reaction as detailed in a lawsuit by the family against the zoo.

The complaint (copies of which you can find online) states, "(the mother) attempted to enter the African wild dog exhibit by climbing through the opening viewing window in the observation deck, but was physically restrained by another zoo visitor. She was forced to watch helplessly as the African wild dogs savagely mauled and literally tore apart her son in front of her."

I can't quite bring myself to imagine the boy's death beyond these surface-level descriptions. That is, I can't bring myself to relive the attack in my mind, to fully appreciate it aesthetically, as Schopenhauer might. Though you can look up the incident and find all the details, it didn't feel right to name the mother and the boy. I have to admit the difficulty for me in approaching the reality of these events. Perhaps it is too close to home, too close to my own children. I can approach those moments just before the boy fell. I can feel the boy's waist in my hands, can feel his anxious energy coiled and ready to spring, his hipbones pushing against my grasp.

Sometimes I hook my fingers through my daughter's pant loops as a kind of extra security measure. I can feel the boy twist, feel my grip fail, my fingers clutching at his pants. I can see him falling. It happens fast. But it's at this moment when my imagination shuts down, when it can no longer follow the arc of the story in my mind. It is at this moment when I recoil from the ecstatic and the sublime.

I can't go there. I can't watch and I have to resort to the facts as I find them or recreate them from newspaper accounts. I like to think I would have leaped into the cage after my son or my daughter, that I would have elbowed the other visitors aside and fought off all eleven dogs to save my child; but perhaps I would have been so paralyzed by the horror, so utterly destroyed in mere seconds, the same seconds it would take for the dogs to kill a child, that I, too, would have stood and watched and hoped that it was all a mistake, just a slip-up, one of those innocent mistakes that any parent can make during a trip to the zoo.

JUAN'S TRIP

On May 19, 1987, three young boys broke into the Prospect Park Zoo in Brooklyn after closing time and found their way to the polar bear enclosure. Perhaps the boys were in search of a thrill, a test of their budding manhood. Perhaps it's simpler than that. I don't know. The boys stripped down to their underwear, took off their shoes, and folded up their clothes, piling them neatly outside the circular pit the bears called home. I can imagine the care with which they performed these actions, and I want to believe it was something their mothers taught them, that they were good boys who did what they were told, even when they were doing things they shouldn't be doing.

Put your shoes on the bottom. Ball your socks and stuff them in your shoes so you don't lose them. Fold your pants on the crease, then in half and again in half. Stack them on your shoes. Fold your shirtsleeves in first, then fold the shirt in half, put it on top. Be neat. Be careful.

I can almost see their small stacks of clothes lined up next to the exhibit. You can see it, too, right? It's there. The facts are just a window that opens into everything else.

Initially I assumed the boys wanted simply to cool off, wanted a break from the often-oppressive heat of a New York summer. Temperatures were only in the upper seventies that day and the heavy grip of summer had yet to descend completely on New York. But even at such temperatures, the humidity in New York can be oppressive, especially if your apartment doesn't have air-conditioning. It might have been cooler that night, but not enough to deter them. Perhaps the boys had just seen the bears' habitat and it looked enticing, the sort of place to have a nice late-night swim. It must have looked like an oasis of sorts. Perhaps they didn't truly understand the danger.

We know that two of the boys went into the polar bears' exhibit. And only one made it out. Eleven-year-old Juan Perez was quickly attacked by one or both of the 900-pound bears and dragged back to the bears' cave as the other boy scrambled to safety.

Siebert, who lived near the Prospect Park Zoo and who might have even crossed paths with these boys at some point, clearly feels both some sense of ownership over the zoo—a proprietary feeling fueled at least in part by nostalgia and by its role as a city zoo and a neighborhood zoo—and he seems to feel some kinship with those boys or at least an understanding of the figurative and metaphorical meaning of their leap. He says of them:

> *City kids in the summer will take the caps off hydrants, anything,*
> *to escape the heat, and if they sneak into a zoo for a swim, there*
> *are a number of less threatening moats—the seal pool, for one—*
> *in which to swim. They, however, sought out not the coldest*
> *pool but the pool of the coldest animal, the animal that stands*
> *for coldness.*

By the time the police and zoo officials arrived on the scene, the bears had eaten most of Juan's legs and were fighting over what little remained of his corpse. Unlike with Scandy and Conrado Mones, these bears were subsequently killed in a hail of shotgun fire, in part because they didn't know if there were other boys still in the cage, and in part because they just didn't know how else to respond to such horror.

PROSPECTING

On September 11, 2014, I walked the mile and a half from my friend's apartment in Flatbush to the Prospect Park Zoo, where I paid a modest $8 for admission and strolled past large metal sculptures of a frog, a fish, and beneath an octopus tunnel canopied with creeping vines, then on into the classic Beaux-Arts central square of a New York City zoo, the same zoo that Charles Siebert eulogized in his essay. The Prospect Park Zoo, like the Central Park Zoo, was part of a massive WPA-funded City Works project directed by Robert Moses, and designed by his longtime collaborator, the architect, Aymar Embury.

When I visited, a sea lion habitat occupied the center square. It was roughly the size of a large fountain or a small swimming pool, with an imitation sea cliff rising up from blue water. Inside the pool, an inner tank was separated from an outer wall by a deep moat, and three California sea lions swam circles, occasionally leaping from the water. One of them repeatedly hopped up on the edge of the pool, balancing over the brink, tipping his head over the moat. A sign said there would be a feeding at 11:30, so I made a note to be back for the show.

Ringing the central square were various zoo offices and a main building that housed several somewhat old-school vitrines of tiny monkeys (marmosets and tamarins; with a larger, stone quarry-like habitat that was home to a troop of hamadryas baboons). When I visited, two tiny baboons chased each other around, fighting over a stalk of celery. It was fun to watch.

On my way out, I stopped a volunteer and asked if she knew where they used to keep the polar bears.

"Oh, they haven't had any animals that big here for a long time, not after we began to understand more about the effect of small habitats on the larger animals."

"Yes, I know. But you don't know where they used to be?"

"No, but some of the old-style cages are now kept off-exhibit and used for storage."

Later, at the sea lion feeding, I would learn that the volunteer's name is Sarah and she contributes her time in the Education Department. Sarah had set up a portable speaker near a large bronze statue of a lioness and her cubs (the only remaining lion at the Prospect Park Zoo), then dragged the microphone chord over to the entrance staircase; and at just before 11:30, a crew of three zookeepers wearing rubber boots and fish buckets on their belts parted the crowd of strollers and entered the Sea Lion Habitat. As Sarah began to talk about the California sea lions, where they came from, their diet and habitat, the zookeepers kept the crowd entertained by making the sea lions do tricks. They slapped high-five, barked, and even did handstands. The kids loved it.

I asked a security guard who'd wandered over to see the show, "Do you know where they used to keep the polar bears here?"

"Yeah, right there," she said, pointing at the sea lion tank. "But they've been gone for a long time."

"Anyone ever jump in there," I asked. "You know, I bet it looks pretty nice on a hot day."

She looked at me a little funny but barely paused before answering.

"Nah," she said, "it's got that big moat around it. I mean, yeah, it would feel nice. But no, nobody ever jumps in."

I think she was wrong about the polar bear cage and pool. I think it was nearby but not right here in the center, but it doesn't really matter. The bears are gone now and only the sea lions remain to tempt and entertain. All the big animals have been moved, "far from our view in a deep, suburban diorama," and the zoo that Siebert holds up as his model for a "city zoo" has been replaced with a "children's zoo," one that, I couldn't help but notice, is pretty accurately attuned to its demographic. Everywhere I looked, there were toddlers toddling and babies in strollers, pushed by parents or grandparents or nannies. I was definitely the odd man out that day.

I thought about those three boys, Juan Perez and his two friends, walking past the zoo on that hot day in 1987, staring through the bars at the sea lion pond and the polar bear pool, those cool blue oases shimmering in the moonlight—staring into what was a very different kind of zoo. As a similarly aged kid in Kansas, on those especially hot and humid nights, my friends and I used to hop fences to swim in the pools at apartment complexes. One of them had these lamps with big white plastic globes, like moons hovering above the fence-line, and if you smacked them hard with your hand, the light would sometimes flicker and fade, dying out and casting the pool into protective darkness.

I can see how the polar bear pool might have called to those boys, how they might have figured the bears for innocent pets, slow and domestic and uninterested in them. I can even see how, as Siebert suggests, they were in search of cold, of some kind of relief and respite from the heat of their days.

Some reports I've read suggest that the move to get rid of cages and to relocate large animals like polar bears or tigers to bigger habitats—a move that radically changed the character of the Prospect

Park Zoo—came about as a response to the deaths of Conrado Mones at the Central Park Zoo and Juan Perez at the Prospect Park Zoo. Juan's death and the death of the bears would forever stain that place and change the way it was both conceived and managed. The deaths of two humans in a cage would ultimately contribute to the death of cages; and zoos would become simultaneously more inviting to the human visitor and less accessible, less intimate, less dangerous, and less edifying.

I looked up at the wrought-iron fence bordering the Prospect Park Zoo, at least twelve feet high, with sharp spears jutting into the sky. It would've been difficult to scale, even for three boys. Perhaps they'd found another way in, a back way, under a fence, or maybe they climbed an overhanging tree and dropped into the zoo. Maybe the fences had been smaller then, easier to climb.

A peacock strutted around the tables where I sat and, cocking one eye in my direction, squawked at me, demanding food or something. I watched as the bird harassed other families, getting too close to their kids, begging for scraps. He was kind of a pain in the ass, peacocking around as he was.

The Prospect Park Zoo felt a little like our zoo in Fresno, though smaller and more intimate, like a city park that housed a few animals. On the day I visited, the employees all seemed bored and waiting for something that would never come.

This probably wasn't a fair assessment, but the meager snack bar and gift shop was completely empty, and the girl working there eyed me suspiciously when I entered. She sat on a ledge near an exterior window, far from her cash register and punched her thumbs into her phone. I could almost hear her thinking, "Don't buy anything. Don't buy anything. Don't buy anything."

The crowds of kids in strollers with their nannies quickly dispersed after the sea lion show, and I, too, wandered off, down a wooded "nature" path called the Discovery Trail, past a habitat that used to contain wallabies and kangaroos, but didn't any longer. They had signs telling you what wasn't there, notes detailing what had been removed or relocated. Across from the empty kangaroo paddock, a small cage held two pacing dingoes, and nearby a couple of red pandas lounged on a lattice of dead trees. I passed alone through the exhibits, mostly unnoticed, walking through an aviary filled with skittish birds, and eventually made my way out, intentionally trying to leave the zoo a different way from how I had entered.

IN THE GARDEN

Not far away, across the Hudson River, in Manhattan, thousands of other people were celebrating the thirteenth anniversary of the attacks on 9/11. Celebrating is not the right word. *Remembering. Grieving. Thinking.* Trying to *contain* it in some way. And if collective thought has a weight, like barometric pressure, it seemed as if I could feel the memories hanging in the air that day like a rain that wouldn't fall; people were quieter, the world a little slower and more patient, perhaps. Or maybe this is just the meaning I stitched to that day in the fabric of my memory.

After the zoo, I walked down and around the block, past the Brooklyn Library with its impressive edifice, and found a café where I could have a beer and a pork sandwich and use the café's free Wi-Fi service to catch up on my facts about Prospect Park.

Many people remember, for example, the "Miracle on the Hudson," on January 9, 2009, when Captain Chesley Sullenberger successfully landed US Airways flight 1549 on the Hudson River after slamming into a flock of Canadian geese shortly after takeoff from LaGuardia Airport.

Not as many people remember that the geese had come from Prospect Park. And fewer people know that "snarge" is the word for the residue left on a plane or in its engine after an encounter with a bird. I didn't know this, or that snarge can clog an airplane engine and, if there's enough snarge, it can disable the engine permanently. Nor did I know that, one year later, in response to the snarge-related

Miracle on the Hudson, federal authorities would authorize the capture and gassing of 1,235 Canadian geese in New York City parks, four hundred of them alone from Prospect Park. Authorities also suffocated 1,739 goose eggs by coating their shells with corn oil.

I sat in that café and read the stories and did the math and thought, *that's a lot of dead birds.* In all 2,974 geese or eggs were exterminated during the program. Next I looked up the official number of dead in the 9/11 attacks: 2,996. A difference of twenty-two. The dead geese alone, if we assume an average weight on the low end of seven pounds, would weigh 8,645 pounds, or almost four-and-a-half tons.

A woman sitting across from me was reading a medical terminology textbook, scribbling notes, and occasionally talked to herself, mumbling the music of her discipline. She existed there, in the bustling middle of this world, and seemed to be studying for a test.

This world is full of tests, I thought, and troubling facts that can't always be calculated, tallied with numbers in the margins. I thought about the weight of grief and the weight of loss and about the other side of miracles. I thought about my children, 2,922 miles away and how, the day before, my son had texted me to tell me that his bus was running late, expecting me to be there for him, and how I had to remind him that I wouldn't be there, that his grandparents would pick him up.

"I'm in New York," I said.

"Oh, right. I forgot," he said, pausing for a moment. "You're just usually here."

I FINISHED MY BEER and my pork sandwich and I walked up the block, crossed the street, and entered the Brooklyn Botanic Garden, a sprawling expanse at least three times the size of the Prospect Park

Zoo, that featured wide grassy meadows, tree-lined paths, a "primitive" path of wood chips that snaked through a dense and wild "local" forest, and a large manicured Japanese garden and koi pond; and everywhere I looked, these tiny plastic plaques on metal sticks poked out of the ground, naming everything. Every shrub, weed, tree, vine, flower, or other foliage was labeled, named, identified, and recorded. It was like walking through a living encyclopedia. Stinkbush. Wild ginger. Red maple and Japanese maple. Tags and signs, Latin names mixed with the colloquial, the landscape pinned with language, a living monument to the human drive to wrestle with chaos, to name things in nature, to control and categorize them.

THAT DAY, THIRTEEN YEARS ago, isn't even a word. We don't have the language to contain the loss. We can barely name it, label it, or control it.

That day is a number. A collection of numbers. Code for a loss we cannot fully calculate.

Numbers dead. Numbers wounded. Numbers gassed or greased. Number of years at war. Numbers saved. My son, thirteen years old now, has never lived in a country that wasn't at war.

That day in 2014, I felt the sinking weight of all the numbers that define us—phone numbers, Social Security numbers, confirmation numbers, account numbers, personal identification numbers, patient numbers, and mileage numbers.

Thirteen years later, I got a little lost in the primitive part of the gardens, chased down a path by a sprinkler; but it was like being lost in a reference book, or on the Internet, researching some subject, surrounded by words, awash in color, too many paths presenting themselves, each one a promise.

ON MY WAY OUT of the tangle, past a wide central lawn, I stopped to sit on a bench. Briefly, reluctantly, the cloudy quilt of sky opened up and rained, but only for a few minutes, just enough to break the grip of humidity over the city. The people picnicking on the lawn had just started to scramble for cover when the rain seemed to rise back up into the clouds and hang in the air like a beaded curtain.

////

NEAR THE JAPANESE GARDEN I stumbled across the "Brooklyn Celebrity Path," a curving walkway of flat stones adorned with the names, engraved in the middle of a leaf pattern, of famous Brooklyn residents; and I traced the path backwards, bouncing from Sandy Koufax to Woody Allen, Norman Mailer, Barbra Streisand, Walt Whitman, Judge Judy, and many other writers, actors, musicians, and celebrities of one kind or another. I snapped a photo of Phillip Lopate's name, the only Brooklyn celebrity I could say I'd actually met; and I found myself looking for one name in particular—Mike Tyson. I thought about Tyson and Treadwell, Stephen Haas and David Villalobos.

I snapped photos of stones and traced the path, but I never found Iron Mike. At the end, I found myself near something called the Shakespeare Garden, which featured herbs, spices, and plants mentioned in Shakespeare's works. Only in Brooklyn, I thought, as I wandered past the Bard's herbs, where I found a nice bench in front of monkshood, garden heliotrope, betony, cardoon, and cascading hopflower.

I sat there for a moment, just enjoying the day, and then I decided to call Corey Sokoler, David Villalobos's attorney. Sokoler had somehow negotiated David's unique plea agreement to avoid the trespassing charge, and he'd said nice things about David to the press. I guess I wanted him to give me a different perspective on the whole thing, a different window into the subject. I wanted to ask him if he thought David was mentally ill or suicidal. I'd left two messages with Mr. Sokoler already, explaining that I just had a few questions and was hoping he might be able to put me in touch with David himself. But so far, he hadn't returned any of my calls.

I opened my notebook, found the number, plugged in my headphones, and dialed. I really didn't expect him to answer, but he did.

"Mr. Sokoler?"

"Yes."

"Hi, I don't mean to bother you but this is Steven Church and I was hoping you might have time to answer a couple of questions," I blurted out, but before I could even finish, he interrupted me.

"No. No, I have no comment on the matter. Thank you. Good-bye," and he hung up.

The beeping noise of his rejection resonated in my headphones. I could hear my own breath. The monkshood and cardoon, the dinosaur kale, even the gardener trimming the evergreen, everything was silent, muffled by the noise in my headphones, drowned out in the absence at the other end of this narrative line. I had nothing. Nobody would talk to me. But I knew the next day I would still try to travel David's path, to walk in his footsteps, if only as a kind of role I was playing in this story.

OUR ZOO

The Fresno zoo's big news the year my daughter turned six was the addition of a king cobra and a Komodo dragon to their decidedly dark and wood-paneled "old-school" reptile house. It's "old-school" to me because with its small glass vitrines, fogged over with moisture, its dark paneled walls and piped-in "jungle" sounds, it reminds me of the "snake windows" in the Natural History Museum at Kansas University, where I spent many hours as a child.

The Fresno zoo does not have a polar bear. And the hippo died long ago. Animals die here all the time. If you spend enough time at the zoo, you begin to see the cycles of death and replacement. They've recently completed a multimillion-dollar renovation of the sea lion habitat: turning a sad fetid pool into a realistic-looking California coastal experience, complete with a massive pool at least ten times the size of their previous home, fake sea cliff "rocks," driftwood, pelicans, and piped-in surf sounds. You can view the sea lions—and one blind harbor seal—from several different viewing stations, including through an underwater window where you watch, through a thick pane, as the sea lions swim and twirl, looping past the glass.

The zoo also boasts the popular Stingray Bay, where guests can pet and feed stingrays that swim around in what looks like a wading pool for children. There are also apparently small toothless sharks in the pool, but you never see them, as they mostly stay huddled at the far edge, away from visitors. A few years ago, nearly all of the rays died mysteriously, but the zoo seems to have recovered from this setback and now hundreds of children stream through there weekly, pawing at the rays and feeding them decapitated minnows. Nearby an Andean condor sits on a dead tree, watching, his wings spread wide like a cape. He looks like an old man dressed up as a vampire for Halloween, trying to scare the children.

The whole place has this weird mix of the macabre and the innocent, the beautiful and the ugly. It is a place of contradiction and odd juxtapositions, and a place of seemingly constant change. In 2014, Fresno voters again passed Measure Z, a bond measure that has funded an ongoing series of expansion projects, updates, and changes at the Chaffee Zoo; and I must admit to feeling a tinge of sadness knowing that my children are outgrowing some of the fun of the zoo.

They recently opened an African Savannah, where lions, rhinos, wildebeest, zebras, and Asian elephants have been added to their collection of giraffes in exactly the kind of new-style "habitat" that Siebert decries, complete with hidden fences to make it seem as if all of the animals are peacefully coexisting in a lush and expansive African savannah. On one of the opening days in the new habitat, a juvenile giraffe ran full speed into some kind of wire or cable and broke its neck. When we visited the next week with my daughter's second-grade class, the guide told us that the other giraffes were "in mourning."

The Fresno Chaffee Zoo is in constant flux, often changing dramatically from one visit to the next or changing over time with repeated visits defined by construction noise and gradual reshaping of the zoo grounds; and it is a place where my daughter sees death and replacement move in surprisingly regular cycles.

One animal passes away, is euthanized or relocated, and another takes its place. Habitats are destroyed and rebuilt, remodeled and reimagined. It happens so much it is no wonder we don't bother to learn the names for most of the animals. They become a series of actors on a collection of stages. When one goes down or quits or moves on, another steps in to keep the show rolling.

MANDY'S TRIP

On Good Friday, 2009, in Berlin, Germany, a recently laid-off teacher, Mandy Knoblauch, climbed over a fence and into the polar bear enclosure at the Berlin Zoo during feeding time. Other zoo visitors stood by and watched her make the leap. Mandy said later that she just wanted to swim in the moat. A nice swim. That's all. Perhaps she also wanted an escape from the crushing depression she felt most days. The woman loved being a teacher and it had been taken away from her. She'd been laid off. Now she wanted to float with the bears. She wanted to be close to something huge and majestic. So she leaped into the abyss. She jumped into the cage. And when she swam close enough one of the bears reached into the water and bit her on the shoulder, trying to drag her up on dry land. But in that ecstatic moment, the woman decided to fight. She pulled free from the bear and swam back across the moat to the wall where zookeepers were already lowering a rope. You can watch the whole incident on YouTube and spot the moment when she realized she didn't want to be eaten, when she realized that there was something entirely too painfully real about this experience. As she tries to escape, the bears go into the water after her, paddling over and clawing and biting at the woman as she's being pulled up. In one terrifying moment as she's hanging there like bait, the rope slips and she drops into the water again; and just as the bears begin to converge on her, one of them biting into her backside, rescuers yank Mandy up and out of their reach.

OUR ZOO

The Fresno Chaffee Zoo was home for thirty years to Betsy the Grizzly Bear, a North American Brown Bear (*Ursus arctos horriblis*), and one of its longest-tenured residents. She was not what you would call intimidating, aggressive, or unpredictable. She didn't have much fight left in her as far as I could tell. And I would be hard-pressed to say that contemplating Betsy was akin to contemplating the sublime or reinvolving myself in the matter of my own existence.

Betsy spent most of her time sleeping at the back of her enclosure, occasionally sprawled out in the dirt. She looked like a pile of fur. I never saw her walk. Not once. Maybe roll around a little bit. She had long yellowed claws that looked in desperate need of a trim, and a dusty shaggy coat of coarse brown hair.

Betsy's habitat was located directly across from the snack bar and it consisted mainly of some dead trees, a small pool of dirty water, rocks, and grass. She was separated from us by a large empty moat, what appeared to be a thin electrified wire, and a short split-rail fence. Because of the location of her habitat, Betsy dealt with a lot of noise. Daily, hourly, she faced hordes of screaming children with ice cream smeared on their faces or corn dogs and giant pretzels clutched in greasy fists.

For a while, two enormous and obnoxious blue macaws were housed in a cage just across from Betsy's habitat, and they would occasionally (but not always) respond to the squawking visitors who congregated outside their cage, screeching, "POLLY WANT A CRACKER!" or "HELLO!"

I don't know how many times I stood there with my daughter, perched at the edge of her habitat, staring at Betsy, waiting for her to move, to do something . . . anything, while other children pointed and screamed or licked feverishly at their ice cream cones, complaining that Betsy wasn't doing anything, that she was just lying there; and I'd think to myself, *I bet she'd do something if I tossed you in there with her.*

MY DAUGHTER AND I spent a lot of time at the Fresno zoo. We felt like veterans or locals. We didn't expect much from Betsy, but we still always paid her a visit to show our respect. She was like the grand-mother of the place. We knew to go weekdays, or early on Saturdays and Sundays, while everyone else was at church. Sometimes it felt like we were the only ones there, alone in the cool mist, my daughter perched on my shoulders, clutching my ears. We moved from cage to cage, peering into the tiny diorama of an actual habitat, hoping to catch an animal playing, walking, or just looking cute.

We lived in an area of Fresno with no public parks or green space. Sometimes we hopped the fence at a nearby elementary school, hoping we didn't get caught. But the zoo was our most consistent escape from the heat—always ten to fifteen degrees cooler (especially in the rainfor-est habitat), and often pretty quiet on those mornings, midweek, when other kids were at school or their parents were working. I could just put my daughter down or push the stroller empty, letting her run free, fol-lowing along behind as I let her lead and guide us into the jungle, the reptile house, past the orangutans and elephants, out into the savannah.

The Malayan tiger, Kiri, always paced incessantly, tracing the perimeter of her cage and carving a dirt path into the grass. As she passed by the viewing window, she made deep huffing sounds like an old woman clearing her throat or grumbling about the food in the cafeteria. A barrier of bamboo separated the tiger's enclosure from the orangutans and siamangs, but the tigers were also caged in on all sides (even the top) with thick wire mesh. Even if you'd wanted to get in there, you couldn't do it. Another split-rail wooden fence separated the casual visitor from the cage, but kids were constantly climbing up on it and leaning over, getting close to the tiger. It would've been easy enough, however, for one of them to stick a hand through the bars, or at least some tempting little fingers.

Kiri the tiger died somewhat mysteriously and unexpectedly in 2011 and was replaced by a male Malayan tiger named Paka, a good "breeder" who'd fathered ten cubs already in captivity. The zoo soon added a female tiger that, in 2014, gave birth to four cubs, fathered by Paka. Visitors could watch via a closed-circuit camera as the mama tiger nursed the cubs. Not long after his cubs were born, however, Paka was euthanized due to what the zoo called "declining health." Now his cubs are the stars of the zoo. But they've grown bigger and will soon be shipped off to other zoos or preserves. The other day when the kids and I visited, I asked one of the zookeepers conducting the "Tiger Talk," if she ever went into the cage with the tigers.

As she shook her head and said, "Nope, never," another keeper piped up, "I like life. I don't want to die," as an explanation for why they never put themselves into direct contact with the tigers.

The keeper, who had two tiny red dot earrings and long curly hair, wore rubber boots and held a pair of barbeque tongs that she used to feed raw hamburger to the tiger.

"Even when the cubs where just three months old and we had to go in to do some blood draws, we wore welding gloves and several long-sleeved layers, and they still ripped through the gloves."

Humans, she explained, just aren't equipped to handle a tiger.

"The cubs used to run up their mother's back with their claws out, bite her on the neck, and then roll off her shoulder. They were just playing. But if they did that to us, we'd be dead."

More recently the zoo has begun to market "behind-the-scenes" experiences that allow visitors beyond the fences. For $35 you can toss fish to the sea lions. For $275 you can spend a day as a zookeeper's apprentice; and all of these experiences are designed not only to make money for the zoo's educational programs but also to allow visitors to get up close and intimate with the animals, and perhaps as Charles Siebert writes, "end up confronting some slightly confining truth about ourselves."

Lyn Myers, an assistant curator at the Chaffee Zoo, said something similar, "When you can get one-on-one with an animal and get close and ask questions and actually be in their environment, I think it brings a whole new idea to learning."

I think David Villalobos would agree. But the Fresno zoo obviously isn't tossing people into the tiger cage, not dropping them into the alligator pond for some wrestling, or letting them try to provoke Betsy the grizzly bear into a fight. Perhaps all that separates Lyn Myers and David Villalobos, though, is a matter of degrees. Villalobos pushed the experience to the furthest end, to the apex of animal encounters. If we learn something unique from proximity to zoo animals, what is it that we learn from proximity to predators? What do we learn from life and death struggles with them? Are the lessons not greater, the consequences more real?

One day not long before Paka died, my daughter and I watched the zookeeper's informational talk on the tigers. Paka paced and chuffed at the fence just behind her as she fed him with the barbeque tongs. She raised the meat up high and the tiger stretched out on the fence like a housecat on a couch—a really big housecat with cartoonishly huge paws and a mouth big enough to squish my head like a grape.

The zookeepers were training Paka to show them his paw, which allowed them to inspect his claws and footpads. But he'd also been trained to stick his tail through the cage wires so the vets could take blood samples. Weighing in at around 260 pounds, Paka was not even one of the biggest tiger breeds, but he was still enormous, still clearly above me on the food chain—even if we did weigh about the same.

The zookeepers always liked Paka because he put on a good show for the guests. Sometimes if you chuffed at him, he would chuff back. Every time I came close to Paka I swear I felt some kind of energy radiating off of him, some wild heat or barely restrained power that the nearby African elephants or the giraffes just didn't have. It was a both terrifying and intoxicating feeling, not unlike being under the influence of a drug.

HER ZOO

My daughter was mostly uninterested in the tigers at the Fresno Zoo. More often than not, if I was staring at Kiri or Paka, watching the zookeeper talk, or trying to find one of the tigers sleeping in his paddock, I'd turn around to find her drifting over to the old African savannah area where her favorite animal, Stoney the Camel, stood in the sun as an ostrich picked bugs from his butt.

Stoney, one of the longest-tenured residents of the zoo, spent his days with the ostrich, the giraffes, gazelles, and elands; and he didn't seem to mind the nitpicking. They appeared to have some kind of symbiotic arrangement. The bird fed off him while Stoney just kept on chewing or staring off into space. He wasn't the most exciting animal at the zoo. Rumor had it that he used to give rides to children at a local park back in the '70s and '80s, and he'd become a sentimental favorite for many of the zookeepers.

I was never sure why my daughter liked Stoney so much, but she definitely went through a serious camel phase. I'm not exaggerating. If anyone asked her, "What's your favorite animal?" her answer, for a while at least, was always, "Camel."

She had several camel stuffed animals and a couple of books featuring camels and llamas. She knew the difference between a Bactrian, two humps, and a dromedary, one hump. Stoney was a Bactrian camel, I think. Like all kids she pretty quickly moved on to other obsessions, but camels held sway for many months, and Stoney was her living model of camel-ness.

Stoney appeared at peace with his life, offering an alternative to the pacing Paka, who emanated angst; and a stark contrast to the barking, howling siamang monkeys nearby, or even his neighbors, the giraffes, who gathered at the fence and licked frantically for leaves at feeding time. And maybe that's what my daughter appreciated about Stoney, that sense of peace, the lack of any anxious energy that seemed to define so many of the other animals in the zoo.

I think that's also part of what I liked about Betsy the grizzly bear. She wasn't interested in performing for visitors, or even moving much at all. I will admit to the briefest of urges to get up close to Betsy, to leap across the moat, into her cage, and see what might happen. It looked like it would be fun to recline on her as if she were a big hairy armchair. I can almost taste the musky smell of her fur. I'm sure, however, that given the right motivation, Betsy could still summon some animal instinct and use those long claws for more than scratching her hide. Her final days were lived out much like the years before them— finding her own peace amidst the human noise surrounding her.

Betsy and Stoney each died within a year of each other, not long before the tiger cubs were born to their neighbors. The zoo didn't get a new camel, but Betsy has now been replaced by a sloth bear that was described to me by a zoo employee as "more social and interactive." The last time we visited, the bear paced circles around a dead tree on a dirt path worried into the grass.

TO TOUCH A TIGER

Conrado Mones seemed unable to help himself, compelled by unseen forces, driven to jump into the cage with Scandy. He may have been mentally ill as reports suggested. Or maybe he simply wanted something he couldn't have, a kind of elusive peace and happiness, something spiritual or sublime. Perhaps this is what Juan Perez and that teacher in Germany and David Villalobos wanted as well.

Perhaps David was not so far from Timothy Treadwell, the conservationist and movie star. Villalobos contested the trespassing charges leveled against him in court, still insisting he'd done nothing wrong. How can you trespass when you're going home? But he eventually pled guilty and was given a discharge on the condition that he complete one year as an inpatient at a mental health facility. It was a strange plea deal, and one where, because of his grand leap into a tiger cage, Villalobos traded what might have been a misdemeanor charge, a fine, and probation for confinement in a cage of his own.

I tracked down a number for Detective McCrossen, who now works homicide for the NYPD, hoping he might be able to confirm a few things and clear up a couple of contradictions I found in the news reports. I wanted to know who'd found David first, what he'd said to whom, and most importantly, whether he had leaped from the tram directly into the tiger enclosure or had climbed a fence to get in, as one report had said. I wanted to know what David acted like in the moment, if he seemed crazy, agitated, pained, or peaceful. I had so many questions.

I reached Detective McCrossen at his desk, and when I mentioned David Villalobos, he chuckled and said, "Yeah, I remember."

I explained that I just wanted to ask him a couple of questions.

"I'll talk with you," he said, "but you gotta go through the DCPI."

"The what?"

"Public Information. You gotta get clearance first before I can talk."

I called and was told I needed to send an email request. That request and a follow-up request, and two more follow-up requests were each ignored. To date I haven't received any response from them. I don't know if my request has just been lost or ignored accidentally, or there's more intention behind the silence.

Most troubling of all for me in this search, though, was the Bronx Zoo's response to my request for clarification. I sent an email and made a follow-up phone call to the Public Relations Department of the Wildlife Conservation Society, the organization that controls the Bronx Zoo as well as the Prospect Park Zoo, where Juan Perez was killed. Call me naïve but I honestly thought it wouldn't be a big deal, that I'd at least get some kind of official report or, if I was lucky, a behind-the-scenes tour. I clearly underestimated their interest in suppressing any stories about the incident and clearly overestimated my journalistic acumen.

After a day or two, I received the following email response from the Director of Communications. He said, "Thank you for your email and voicemail today. We are going to pass on the opportunity to participate."

Pass on the opportunity to participate.

Those words. I could hardly understand them, though I appreciated their euphemistic glow. This whole search for me was a way to *participate* in the story. It was all because I couldn't choose not to *participate*, because the story wouldn't leave me alone. I couldn't understand how or why anyone would *pass on the opportunity to participate.* Besides that, the zoo had already participated in the story, whether they liked it or not.

I started to wonder if there was something dangerous about my questions, something that might expose them to unwanted criticism or attention. Perhaps my question about whether Villalobos had leaped directly into the tiger enclosure, or whether he'd jumped down and climbed over a fence, exposed a design flaw and a safety issue for the zoo.

Maybe it was an insurance liability and an open wound, an ugly and aberrant event that they just didn't want to acknowledge. Maybe they just didn't want to participate in my personal essaying of these events. Maybe they thought any attention to the story would only encourage copycats. In the absence of their participation, I had only my own experience and imagination to explore the deeper truths of David's leap.

This place will always belong to him. This zoo. This island of trees and hills, this green space of thought. No matter how many thousands or millions of visitors. David claimed it. Planted a flag with his leap; and nobody wants to admit this. They want to believe he never landed, that he leaped into the ether and disappeared into history. Perhaps David and Timothy Treadwell and Stephen Haas, Conrado, Juan, and Mandy, all wanted to be closer to death—but they wanted this because it's only in such fleeting moments where immortality exists, where you are most fully and completely and terrifyingly alive. In such liminal spaces, however brief, anything is possible.

THE POETRY OF A LEAP

Melissa (name changed to protect her identity) was eating a sandwich when I approached her at the Bronx Zoo tram station on that warm September day in 2014. She wore a navy blue windbreaker over her zoo uniform shirt, shorts, and had her hair pulled back in a tight ponytail. I said hello and asked how her day was going, and she told me things had been kind of slow. We both watched a group of British tourists marvel and squeal at the sight of a black squirrel snoozing in a tree.

"Hey," I asked, seizing my opportunity, "is it true that some guy jumped from that monorail into the tiger cage?"

Melissa looked over her glasses and up at me for what seemed like a really long time. She took a bite of her sandwich and studied me.

"I'm not allowed to talk about it," she said, chewing her food, "but yeah, a guy jumped."

She wiped her face with the back of her hand and didn't take her eyes off of me.

"Why can't you talk about it?" I pressed.

"I could lose my job."

"Is it some kind of legal thing?"

"Yeah, you know we get like the paparazzi in here and they want to dig up the details."

"And you're not allowed to talk about it."

"No. No, I'm not."

I climbed aboard the tram, heading for the bears on the other side of the park, and I tried to imagine myself as "paparazzi," a word that to me always sounded so much more beautiful than what it actually signifies, as if it describes some kind of Italian moth drawn to the flame of a story. And I suppose I *was* trying to dig up details, hoping to excavate something from my trip to the zoo, but I didn't think I was chasing the sensational and horrific. I didn't even have a camera and I wasn't taking many notes. I was more interested in the mundane details behind the sensational story, and certainly wasn't interested in getting anyone fired; but it was still a little hard for me to understand why nobody wanted to talk about what had happened. The stubborn refusal of everyone to talk made it all the more mysterious for me. Were they hiding something? Probably not. But the point was that I'd never know, never would hear the story of David's leap from their perspective. All I could do was ride the tram and try to imagine that day, try to see what he saw.

A couple of hours later, I stood in line for my second trip on the Bengali Express Monorail for "research purposes," and Melissa saw me there, clutching my ticket. She'd obviously left her spot at the tram station and moved over to work the monorail for part of the day. I waved and she recognized me, so I just smiled and nodded in acknowledgment; and there passed between us some kind of moment, a quivering couple of seconds where we entered a different relationship together, something darker than familiarity.

"I'm scared of you," she said across the turnstiles.

"What?" I said, confused at first, thinking she meant that stuff about paparazzi.

But then the moment was gone, twisted to a joke as she chuckled and turned away, talking to her coworkers, ignoring me. And I knew suddenly what she meant. It hit me. She thought I was another David. She wasn't worried about me being paparazzi. She thought I might jump.

Don't be scared, I wanted to say. *That's not my story.*

Deep down, of course, I do think many of us at least want to reach our hand through the cage bars and to try to touch something beyond our control, something alien and animal and alive. It's like trying to touch the absence of you, the certainty of death, your own annihilation. It's an inversion of power, a hand reaching across the line between predator and prey. And some of us, it seems, want to do more than touch. Some of us want to leap over the line and embrace it, want to jump into the cage; and maybe, just maybe what ultimately separates us from the animals in the zoo is not only the physical cage but also this pathological drive to turn something natural and animal—savagery and death—into something spiritual, meaningful, and utterly human. Any meaning that extends beyond the moment is ultimately a fragile construct.

I didn't want to jump.

Just as I learned when I became the animal and slammed that guy against the wall, in real life there is rarely narrative resolution, or an arc of redemption or salvation that makes the violence meaningful. It's just natural, not magical or mystical. Timothy Treadwell may have died believing the opposite, believing that he had finally become "one with the bears," as crass and insensitive as that may sound; but I think I'm still in the Werner Herzog camp, still believing that a great predator like a bear or a tiger meets our desire to imbue meaning, metaphor, or morality with little thought beyond total indifference.

I didn't want to jump.

A grizzly bear cares little for your love. A tiger isn't interested in the poetry of your leap. A polar bear will ignore the art of your therapeutic swim. Once inside the cage, to them you are just meat, intruder, trespasser, and toy. You are prey. You ARE animal. All the barriers we put up to insulate us from this fact fall away in the moment.

The cage—still a stage, but for a different kind of story, a more elemental tale—no longer protects you from reality. And, just as it should be, no amount of prayer or peace is going to protect you from the inevitable end unless, like David, you somehow encounter an animal with more mercy, self-control, and patience than you can give as a sacrament, as an offering to the other side.

ORIGIN STORIES

The savage in man is never quite eradicated.

— HENRY DAVID THOREAU

My mother long ago stopped asking "why?" when it came to my writing projects; and when I told her I was researching animal attacks, she asked me if I remembered our trip to the circus in Topeka, Kansas, in 1977.

"No, I don't remember anything," I confessed.

"Really? I'm surprised. It seems so much a part of what you're writing."

Mom told me we'd been sitting close to the lions' cage, and in the middle of the big cat show featuring a young lion-tamer and performer named Josip Marcan. The show had started and everything seemed normal, when one lion lunged suddenly forward and bit Marcan on the arm, then swatted at his face and head with his paw. Other lions soon joined in the attack, pulling against their chains, roaring and lunging as Marcan tried to gain control of the beasts; but we didn't see how it ended or what happened.

Mom and Dad hustled my brother and me up from our seats and out of the arena.

"I don't think we even talked about it," she said.

We just climbed in the car and drove home. We are, after all, Midwestern at our core. We didn't talk about the bad stuff. If we didn't talk about it, didn't name it, then it wasn't real. I had no memory of the event, perhaps because we never constructed a story around it, never talked about it during family dinners. We never said, "Remember that time we saw the lion attack that man?" If we had, I wonder how the story might have changed over time.

What Mom remembers most now, though, is the noise and the way it echoed all the way out into the halls of Memorial Auditorium. All the other lions and tigers began roaring and yowling, too, a chorus of them bellowing for more—more blood, more vengeance or justice or food.

"It was haunting," she told me. "I still remember the sound."

I was six years old and just coming out of a long stretch of child-hood illness; perhaps my brain hadn't fully recovered from my febrile seizures and wasn't cataloguing memories the way it should. I cannot even conjure the faintest impression of this experience. It's true that most of my memories before the age of seven or eight are fuzzy with fiction, hazy and out of focus; and no matter how much I twist the eyepiece, they never come clear. This is one I'd like to see again, if only through the lens of imagination.

Though I can't claim any clear memory of the lion attack on Marcan, it seems likely that it imprinted itself on my young mind and had some influence on my lifelong interest in animal attack stories, particularly those stories where it seems like the human had it coming, where a man had leaped into the cage and asked for such intimacy, ecstasy, and violence.

////

THE YEAR IS **1977.** The year is 1987. The year is 2007. The year is 2012. The year is 2014. And 2015. And 2016. The years. The stories don't stop. They're both real and unreal, pop culture tropes, tiny little windows in the archetypal diaspora.

A family visiting an aquarium at Disneyland is surprised to find Mom swimming with the fishes. She's blowing bubbles and twirling in the seaweed. Her face shines behind the glass like a strange moon. They are selling immersion in this television ad. They are selling an experience.

A three-year-old boy falls into a jaguar enclosure in Arkansas. He is immediately set upon by two big cats. Zookeepers chase them back with fire extinguishers and rescue the boy, but not before he is severely mauled.

A zookeeper dangles meat in front of the lions. He tosses one steak and teases a cat with the other. A mother covers her child's eyes. He is selling something for the Ford motor company. I don't remember what. Security, perhaps. Peace of mind. The archetype persists.

A woman and her driver, visiting a wildlife preserve in South Africa, approach a pride of lions, taking photographs with their car windows down. A lioness lunges through the open window, killing the woman and maiming the driver.

Fonzie jumps the shark tank on his motorcycle. Nobody believes he won't make it. Nobody is going to let Fonzie get eaten by sharks.

But part of me wanted him to fail. I at least wanted to believe it could happen. That he would fall into the abyss. That he would be consumed. That his iconic leather jacket would float, disembodied, to the surface.

A young man falls or jumps—it isn't clear—into a tiger enclosure in India. He squats, curled up tight against the fence, as if he's trying to disappear. But the tiger sees him. The tiger isn't going to let him leave.

A man on the street provokes Mike Tyson into a fistfight. He pokes the tiger with a stick. He pushes Tyson's buttons. Who does that? Who asks a tiger for a fight?

A man in Berlin leaps into the brown bear habitat. He wants to dance. The bear grabs him by the shoulder and drags him around for a while. You can see the fun fade and the pain rise up in his face, as his shirt is painted red with his own blood.

John Jeremiah Sullivan publishes his essay, "Violence of the Lambs," a satirical article on the rise of animal attacks and the possibility that they're going to wipe out humans and take over the planet. It is, in many ways, a very convincing essay.

A woman in Thailand leaps into a pit of alligators and dies horrifically. They rip her apart. Of course. Because that is what she wanted them to do. That was the plan. And part of me appreciates the execution of her plan.

////

THESE LEAPS. THESE TESTS. These advertisements and clichés. These spiritual encounters. These plagues of beasts. These *zooicides*. These stories. They scroll across my newsfeed, my television, my phone, my consciousness. People send me links and quotes. They know that I am a moth to the flame. Perhaps I *am* paparazzi. Say hey did you see this. Say hey did you hear about the boy who fell into the jaguar pit, the boy who fell into the gorilla habitat. Say hey did you hear about the boy and the tiger, the boy and the alligator. Say hey did you hear about the boy and the bear, the man and the bear, the woman and the bear. And I've come to understand that David Villalobos is as much an archetype as anything. He is Daniel, a classic character in an ancient story, a tale retold over and over again. The tragic figure cast into the pit, facing the lions, the tigers, the bears. But now the story often drifts into the absurd.

We create and sustain such archetypes because they speak to some universal truth about human life, because the test of the leap is a test we all understand; but what do we do with archetypes that we don't want to talk about? How do we handle the taboo as archetype, the forbidden but seductive role of intentional victim, the man or boy or woman who leaps into a cage with an apex predator? What do we make of a willing Daniel who throws *himself* into the lion's den, he of a savage mind who calls such violence down upon himself?

Mostly, I think we call them aberrations, anomalies, freaks, outcasts, mentally ill, or suicidal. I think we try to define their leaps, name them and control them. But we cannot always contain the reach and depth, the viral spread of their stories. We can't really understand. We can't even come close.

David Villalobos, vessel and archetype, character and caricature, charismatic barbarian, has become like Stephen Haas to me, both real and unreal, an amalgamation and composite, a creation of my own mind as much as a citizen of the real world. He contains contradictions: hero and victim, protagonist and antagonist, perhaps even man and savage. I don't mean to diminish David's actual existence. I know he is greater than I can possibly imagine.

I can see the edges of the puzzle picture bleeding out, fading into the surrounding noise; and I understand that no matter what his leap meant to David at the time, it will never mean the same thing again. I also don't want to reopen old embarrassing wounds, or victimize David by appropriating parts of his story—though I realize that I've perhaps already done this.

I want for David Villalobos, I suppose, what I want for all charismatic barbarians, for all the manimals of the world—what I want for Stephen Haas and Treadwell and Mike Tyson, and even for myself— to *humanize* the choice to leap, to face the savage and the wild inside. I want to acknowledge and accept that elemental desire to dwell in this hazy boundary between human and animal, a liminal space that perhaps defines itself most sharply in moments of ecstatic violence and savagery.

I suppose I want to give a heart to the barbarian, and to normalize the need to get intimately close to apex predators, so close even that you become one with the tiger.

SOURCES CONSULTED

All quotes at the beginning of each part that reference "savagery," and most quotes attributed to Mike Tyson, were found by doing searches in a variety of online quote databases such as the following:

www.brainyquote.com/quotes/quotes/o/octaviabut322227.html

www.quotehd.com/quotes/author/mike-tyson-boxer-quotes

PROLOGUE: DAVID'S LEAP

Esposito, Richard, and Colleen Curry. "Man Mauled by Tiger at Bronx Zoo Charged with Trespassing." ABC News Network, Sep. 22, 2012. http://abcnews.go.com/US/man-mauled-tiger-bronx-zoo-charged-trespassing/story?id=17296069

"Horror at New York Zoo as Man, 25, Is Brutally Mauled by a TIGER after Leaping into Its Den from Monorail in Bizarre Suicide Attempt." *Mail Online*. Sep. 22, 2012. www.dailymail.co.uk/news/article-2206932/David-Villalobos-mauling-Horror-New-York-zoo-man-25-brutally-mauled-TIGER-leaping-den-monorail-bizarre-suicide-attempt.html

Lestch, Corinne. "Man Mauled by Bronx Zoo Tiger Pleads Not Guilty." *NY Daily News*, January 11, 2013. www.nydailynews.com/new-york/bronx/man-mauled-bronx-zoo-tiger-pleads-not-guilty-article-1.1238201

MTV Movie Awards Sketch. "When Leo Got Fucked by a Bear." 2016. www.youtube.com/watch?v=AyQ9rW5ntsk

The Revenant. Dir. Alejandro Iñárritu. Perf. Leonardo DiCaprio, Tom Hardy, Will Poulter. Regency Enterprises, 2015. Film.

"Wild Asia Monorail." Bronx Zoo. N.d. http://webmail.bronxzoo.org/animals-and-exhibits/exhibits/wild-asia-monorail.aspx

PART ONE: STEPHEN HAAS

"Bear Kills Woman and Her Son in Alaska." *New York Times*, July 4, 1995. www.nytimes.com/1995/07/04/us/bear-kills-woman-and-her-son-in-alaska.html

George, Jean Craighead. *My Side of the Mountain*. New York: E.P. Dutton, 1959. Print.

Glacier Park's Night of the Grizzlies. Montana PBS. 2010. Documentary. www.montanapbs.org/GlacierParksNightoftheGrizzlies/

Jabin, Clyde. "Grizzlies Turn Killer—Two Girls Die, Youth Mauled." *Windsor Star*, Aug. 14, 1967, p. 1. Windsor Star - Google News Archive Search.

Krakauer, Jon. "Death of an Innocent: How Christopher McCandless lost his way in the wilds." *Outside Magazine*, Jan. 1993.

"List of Fatal Bear Attacks in America." Wikipedia. Last modified June 11, 2016. https://en.wikipedia.org/wiki/List_of_fatal_bear_attacks_in_North_America

McKay, Brett, and Kate McKay. "Bear Attack Survival Guide." The Art of Manliness. 2008. www.artofmanliness.com/2008/01/30/how-to-survive-a-bear-attack

"Online Etymology Dictionary." www.etymonline.com. N.d.

PART TWO: TIMOTHY TREADWELL

Bissell, Tom. "The Secret Mainstream: Contemplating the Mirages of Werner Herzog." *Harper's Magazine*, Dec. 2006.

Burke, Kerry, Kerry Wills, and Barry Paddock. "ZOO-ICIDE: Man mauled after leaping 17 feet into Bronx Zoo tiger den in crazed bid to kill himself." *New York Daily News*, Sep. 12, 2012. www.nydailynews.com/new-york/bronx/man-injured-jumping-falling-tiger-den-bronx-zoo-article-1.1165004

Farley, Frank. "The Type T Personality." In L.P. Lipsett and L.L Mitnick (eds), *Self-Regulatory Behavior and Risk Taking: Causes and Consequences*. Norwood, NJ: Ablex Publishers, 1991.

Goleman, Daniel. "Why Do People Crave the Experience?" *New York Times*, Aug. 2, 1988. www.nytimes.com/1988/08/02/science/why-do-people-crave-the-experience.html

Grizzly Man. Dir. Werner Herzog. Perf. Werner Herzog, Timothy Treadwell. Lionsgate Films, 2005. Film.

The Holy Bible. Daniel 6:22, "and he shut the mouths of the lions."

Munsey, Christopher. "Frisky, but more risky." *Monitor on Psychology*, July/Aug. 2006, Vol. 37, No. 7, www.apa.org/monitor/julaug06/frisky.aspx

"Schopenhauer's Aesthetics." Stanford Encyclopedia of Philosophy. http://plato.stanford.edu/entries/schopenhauer-aesthetics/#Sub

Zuckerman, Marvin. "Chapter 31. Sensation seeking." In Leary, Mark R. and Rick H. Hoyle, *Handbook of Individual Differences in Social Behavior*. New York/London: The Guildford Press, 2009, pp. 455–465.

PART THREE: CHARISMATIC BARBARIANS

George, Carmen. "Man arrested after allegedly touching orangutan at Fresno Chaffee Zoo." *Fresno Bee*, April 21, 2015.

Hittel, Theodore. *The Adventures of James Capen Adams, Mountaineer and Grizzly Bear Hunter of California*. 1860.

Hulk. Dir. Ang Lee. Perf. Eric Bana, Jennifer Connelly, Sam Elliot, Nick Nolte. Universal Pictures, 2003. Film.

Sellier, Charles. *The Life and Times of Grizzly Adams*. New York: New American Library, 1977.

Shenk, Dean. "Yosemite Nature Notes." *Yosemite National Park*, Vol. 45, No. 2, April 1976.

PART FOUR: THE ANIMAL WITHIN

Friedman, Myra. "My Neighbor Bernie Goetz." *New York*, Feb. 18, 1985.

Hornblower, Margot. "Intended to Gouge Eye of Teen, Goetz Tape Says; 'My Problem Was I Ran Out of Bullets," *Washington Post*, May 14, 1987.

Hulk. Dir. Ang Lee. Perf. Eric Bana, Jennifer Connelly, Sam Elliot, Nick Nolte. Universal Pictures, 2003. Film.

"Human Aggression." *Annual Review of Psychology*, Vol. 53: 27-51, February 2002.

Leaf, Munro (Author); Lawson, Robert (Illustrator). *The Story of Ferdinand.* New York: The Viking Press, 1936.

"'You Have to Think in a Cold-Blooded Way'" [transcript of Bernhard H. Goetz police interview]. *New York Times*, April 30, 1987.

PART FIVE: IRON MIKE
Blue Velvet. Dir. David Lynch. Perf. Dennis Hopper, Kyle McLachlan, Isabella Rosselini. DeLaurentis Entertainment Group, 1986. Film.

Manger, Warren. "'Sex Addiction' Stopped Mike Tyson from Being the Best Boxer Ever." *Mirror.* Feb. 8, 2015. www.mirror.co.uk/sport/boxing/sex-addiction-stopped-mike-tyson-5128752

Mann, Tedd. "Cop who shot marauding chimp says workers' comp law a 'farce'." *The Day*, Feb. 22, 2010. www.theday.com/article/20100226/NWS12/302269906

"Mike Tyson's Famous Interview and Quote." YouTube. 2008. www.youtube.com/watch?v=FWeD5KXx5WI

Oates, Joyce Carol. *On Boxing.* Garden City, NY: Dolphin/Doubleday, 1987.

Reservoir Dogs. Dir. Quentin Tarrantino. Perf. Harvey Keitel, Tim Roth, Michael Madsen. Fyodor Productions, 2008. Film.

Robertson, Nan. "The All-American Guy Behind *Blue Velvet.*" *New York Times*, Oct. 11, 1986.

Sandoval, Edgar, and Rich Schapiro. "She Lost Eyes, Nose and Jaw to Crazed Chimpanzee." *NY Daily News*, Feb. 19 2009. www.nydailynews.com/news/charla-nash-lost-eyes-nose-jaw-chimpanzee-attack-article-1.365935

"Stuck in the Middle with You." *Stealers Wheel.* A&M Records, 1972. LP.

Tyson. Dir. James Toback. Perf. Mike Tyson, Mills Lane, Trevor Berbick. Fyodor Productions, 2008. Film.

Wilson, Michael. "After Shooting Chimp, a Police Officer's Descent." *New York Times*, Feb. 24, 2010. www.nytimes.com/2010/02/25/nyregion/25chimp.html?_r=0

PART SIX: FATHER AND DAUGHTER

Badger, T.A. "Alaska's Mr. Popularity: Binky the Bear." *Seattle Times*, Oct. 2, 1994. http://community.seattletimes.nwsource.com/archive/?date=199 41002&slug=1933698

Barron, James. "Polar Bears Kill a Child at Prospect Park Zoo." *New York Times*, May 20, 1987. www.nytimes.com/1987/05/20/nyregion/polar-bears-kill-a-child-at-prospect-park-zoo.html

"Bronx Zoo." *NYC Parks*. www.nycgovparks.org/about/history/zoos/bronx-zoo

"Celebrity Path." Brooklyn Botanic Garden. www.bbg.org/collections/gardens/celebrity_path

Ellis, Richard. *On Thin Ice: The Changing World of the Polar Bear*. New York: Alfred A. Knopf, 2009.

Golgowski, Nina. "Mother of a toddler fatally mauled by African dogs is to blame and shouldn't be allowed to sue: zoo." *New York Daily News*. Sept. 12, 2013. www.nydailynews.com/news/national/mom-blame-boy-mauled-death-zoo-article-1.1453399#ixzz2j52dhzdr

"Man Killed in Bear Cage Was an Immigrant." *Gainesville Sun*, Sep. 28, 1982.

"Man Polar Bear Killed Identified as a Cuban." *New York Times*, Sep. 28, 1982.

McFadden, Robert D. "Pilot Is Hailed After Jetliner's Icy Plunge." *New York Times*, Jan. 15 2009. www.nytimes.com/2009/01/16/nyregion/16crash.html?_r=0

Saucedo, Carlos. "Fresno Chaffee Zoo Gives Visitors an Up Close Experience with Animals," ABC30 Action News. Sep. 24, 2013. http://abclocal.go.com/kfsn/story?section=news/local&id=9259439

Shipp, E.R. "Polar Bear in the Central Park Zoo Kills Man Who Climbed into Cage." *New York Times*, Sep. 27, 1982. www.nytimes.com/1982/09/27/nyregion/polar-bear-in-the-central-park-zoo-kills-man-who-climbed-into-cage.html

Siebert, Charles. "Where Have All the Animals Gone? The Lamentable Extinction of Zoos." *Harper's Magazine*, May 1991. www.harpers.org/archive/1991/05/where-have-all-the-animals-gone-the-lamentable-extinction-of-zoos/

"Toledo youth jumped into bear pit after taking drugs earlier." *Bryan Times*, Jan. 25, 1972. https://news.google.com/newspapers?nid=799&dat=19720 125&id=dk8LAAAAIBAJ&sjid=a1IDAAAAIBAJ&pg=7324,1315737& hl=en

"Woman Attacked by Polar Bear after Jumping into Tank at Berlin Zoo." YouTube. 2009. www.youtube.com/watch?v=wKDu1qHLv4E

EPILOGUE: ORIGIN STORIES

Barry, Ellen, and Nida Najar. "White Tiger Kills Visitor at Zoo in India." *New York Times*, Sep. 23, 2014. www.nytimes.com/2014/09/24/world/ asia/white-tiger-kills-visitor-to-new-delhi-zoo.html

"Boy, 3, injured after falling into Arkansas zoo's jaguar exhibit." Fox News. Oct. 12, 2014. www.foxnews.com/us/2014/10/12/boy-3-injured-after-falling-into-arkansas-zoo-jaguar-exhibit.html

Findlay, Stephanie. "US woman dragged from car and killed by lion in South Africa park." *Telegraph*, June 1, 2015. www.telegraph.co.uk/news/world-news/africaandindianocean/southafrica/11643991/US-woman-dragged-from-car-and-killed-by-lion-in-South-Africa-park.html

Hall, Allen. "Man mauled after trying to have picnic with bear in Swiss zoo." *Telegraph*, Nov. 25, 2009. www.telegraph.co.uk/news/newstopics/ howaboutthat/6651144/Man-mauled-after-trying-to-have-picnic-with-bear-in-Swiss-zoo.html

"Show Fire STAR INJURED." Circus Historical Society. March 14, 1977. www.circushistory.org/Publications/CircusReport14Mar1977.pdf

Sullivan, John Jeremiah. "Violence of the Lambs," *GQ*. Oct. 13, 2011. www. gq.com/story/john-jeremiah-sullivan-violence-lambs-future-human-race

"Thailand woman dead after jumping into crocodile pit." BBC News Online. Sep. 16, 2014. www.bbc.com/news/world-asia-29218557

"Tigers of India Josip Marcan." 911 Animal Abuse. 2015. www.911animalabuse. com/tigers-of-india/.

NOTES

PART ONE: STEPHEN HAAS

- Many thanks to Jefferson Beavers, author of the original Stephen Haas bear attack story and class assignment, for inviting me to participate in the mock interview.
- It should be noted (even if it seems pretty obvious) that the italicized interview questions and responses from the fictional Stephen Haas in this chapter are entirely imagined and fabricated but also based on my personal experience participating in the mock interview. Because I had no record of this "interview," I worked to try and recreate the essence of it here while also characterizing Stephen Haas a bit more.
- The "List of Fatal Bear Attacks in America," culled from various sources, proved to be an irresistible rabbit hole of research: https://en.wikipedia.org/wiki/List_of_fatal_bear_attacks_in_North_America

PART TWO: TIMOTHY TREADWELL

- Most quotes from Timothy Treadwell, Werner Herzog, and others in this section are taken from the full-text script version of the film *Grizzly Man* or from direct transcriptions I made while watching the film over and over again.
- Sections 4, 5, 6, and 7 reference several articles on research conducted into the psychology of thrill-seeking by Seymour Epstein, Marvin Zuckerman, and Frank Farley.

PART THREE: CHARISMATIC BARBARIANS

- Most references in this section to specific TV shows and movies, including plot summaries and promotional materials, were made through liberal use of www.imdb.com and by watching many, many clips on www.youtube.com.

- Though I was able to find numerous references to a 1978 interview with Charles Sellier in *TV Guide*, the closest I could come to finding the actual interview seemed to come in response to a question on a *TV Guide* forum: www.tvguide.com/news/question-66694/

- If you haven't already, I encourage you to watch the amazing Manimal "panther transformation" from YouTube: www.youtube. com/watch?v=-HhOoZRWYOU

- The section "Smell the Flowers" makes reference to an account of the Bear vs. Bull fights in California. This refers to a possibly fabricated, certainly embellished first-hand account of life in California during the mid-nineteenth century, written by Hinton R. Helper and self-published in 1855. "The Land of Gold: Reality Versus Fiction" is composed as a series of epistolary reports "back home" on the "truthfulness and falsehoods" about a place that had already gained a mythic reputation. Helper, it seems, meant to disabuse his readers of their ideas about California. The book chronicles Helper's journeys through the state, "afford(ing) me ample time and opportunity to become too thoroughly conversant with its rottenness and corruption, its squalor and its misery, its crime and its shame, its gold and its dross." The book—never intended to be a book, which Helper admits in his preface—has ended up being cited by several other historical accounts of life in nineteenth-century California.

PART FIVE: IRON MIKE

- A great deal of this chapter was written by referencing YouTube videos of Tyson's infamous fight with Evander Holyfield (www.youtube.com/watch?v=WozBtwREvFk). It was also influenced heavily by the 2008 documentary *Tyson* and Joyce Carol Oates's seminal work of fight literature, *On Boxing.*

- Per the IMDB.com movie database, Madsen's character is identified as "Mr. Blonde" rather than the masculine form, "Blond."

ACKNOWLEDGMENTS

This book has been a long strange journey; and I've had more fun than is probably appropriate given the subject matter. I owe a great deal to many people for helping shepherd this book to publication. First and foremost, I want to thank Dan Smetanka, who believed in the book from day one, and everyone at Soft Skull Press. I also want to thank my wife, Andrea Mele, and my children, Sophie and Malcolm. Thanks also to my parents, Sally and John Ramage, Ed and Carolyn Church; my brother, Cory Church; Jim and Debbie Mele, and my extended family. Many thanks to my colleagues and students at Fresno State in the MFA Program, the Department of English, and the College of Humanities, and to my coeditors and friends, Matt Roberts, Sophie Beck, and Adam Braver; and massive thanks to an extended network of friends, writers, and editors who've been supportive of this project: Rob Shufelt, Justin Hocking, Elena Passarello, Jill Talbot, Matthew Gavin Frank, Hattie Fletcher, Dinty W. Moore, Jeffery Gleaves, Kristen Radtke, Taylor Goldsmith, Jefferson Beavers, and others too many to name, including my amazing students at Fresno State. Thanks to Taylor Petersen for her awesome research and copyediting assistance. Thanks also to *Creative Nonfiction* and their imprint InFact Books, to *Salon.com*, *Prairie Schooner*, the *Brevity* blog, and *The Nervous Breakdown,* where portions of this manuscript originally appeared. I also want to extend special gratitude to James and Coke Hallowell for their continued support of the Hallowell Professorship in Creative Writing, which has allowed me some much-needed time to work on this book. Finally, I want to say thank you to David Villalobos, the inspiration for this project, and to all the other manimals whose stories have helped shape this journey.

ABOUT THE AUTHOR

S teven Church is the author of *The Guinness Book of Me*, *Theoretical Killings*, *The Day After 'The Day After'*, and *Ultrasonic*. His essays have been published and anthologized widely, including in *Best American Essays* and most recently in *After Montaigne: Contemporary Essayists Cover the Essays*. He's a founding editor and nonfiction editor for *The Normal School*, and he teaches in the MFA Program at Fresno State, where he is the Hallowell Professor of Creative Writing.